Thomas Everitt Coleman

Sanitary House Drainage

It's Principles and Practice

Thomas Everitt Coleman

Sanitary House Drainage
It's Principles and Practice

ISBN/EAN: 9783743687158

Printed in Europe, USA, Canada, Australia, Japan

Cover: Foto ©berggeist007 / pixelio.de

More available books at **www.hansebooks.com**

SANITARY HOUSE DRAINAGE,

ITS PRINCIPLES AND PRACTICE.

SANITARY HOUSE DRAINAGE,

ITS PRINCIPLES AND PRACTICE.

A HANDBOOK FOR THE USE OF
ARCHITECTS, ENGINEERS, AND BUILDERS.

BY

T. E. COLEMAN,

SURVEYOR, ROYAL ENGINEER CIVIL STAFF;
MEMBER OF THE SOCIETY OF ARCHITECTS AND MEMBER OF THE
SANITARY INSTITUTE.

WITH NUMEROUS ILLUSTRATIONS.

London:
E. & F. N. SPON, 125 STRAND.

New York:
SPON & CHAMBERLAIN, 12 CORTLANDT STREET.

1896.

PREFACE.

A SENSE of the vital importance appertaining to the proper construction of every system of house drainage has prompted me to set forth, in a concise form, the essential sanitary principles which should govern the general design of the domestic drainage scheme as a whole, and also the construction of its component details. I have at the same time given careful consideration to the most satisfactory means which have, up to the present, been devised to carry those principles into practice, illustrating the various methods by explanatory sketches.

The matter comprised within the following pages recently appeared in a series of articles in *The Building News*, and as now collected and arranged it will, I trust, form a convenient and practical handbook on the subject of house drainage, not only for students, but also for those who are actively engaged in designing, superintending, or carrying out works of this description.

T. E. COLEMAN.

LINDEN GROVE, GOSPORT:
   September 1896.

CONTENTS.

CHAPTER I.

PRELIMINARY REMARKS. GENERAL PRINCIPLES.

PAGE

PRELIMINARY REMARKS:—Necessity of a sanitary drainage system for the maintenance of health—Drains often laid with insufficient supervision—Attention to minor details an important matter—Explanation of the "conservancy" and "water-carriage" systems of drainage—The "water-carriage" system in general use 1

GENERAL PRINCIPLES:—Definition of "sewage" and "storm-water" drainage—Object to be attained in every system of house drainage—Water supply to be distinct from drains—Drains, &c., to be placed outside buildings—Precautions to be adopted where this cannot be carried out—House drainage to be disconnected from public sewers—Purpose of disconnection—Thorough ventilation of drains necessary 4

CHAPTER II.

SYSTEM OF DRAINAGE. STORM-WATER SECTION. FOUL-DRAINAGE SECTION.

SYSTEM OF DRAINAGE:—Objections to storm-water gullies connected directly to foul drains—Foul and storm-water drains to be grouped into distinct sections—Disposal of storm water—Disconnection of storm drains from foul drains—Distribution of drainage under the separate system 6

STORM-WATER SECTION:—Rain-water only admissible—Gullies to be trapless—Intercepting chamber required at junction

	PAGE
with foul-drainage section—Free ventilation to be provided through all storm-water drains..	8
FOUL-DRAINAGE SECTION:—All gullies to be trapped—Ventilation of drains—Precautions to be observed respecting internal fittings—Disconnection from public sewer	8

CHAPTER III.

GENERAL PLAN. VOLUME OF SEWAGE AND STORM-WATER. GRADIENTS OF DRAINS.

GENERAL PLAN:—Its preparation—Arrangement of drains—Inspection chambers—Foul air outlets—Stable drainage to be distinct from house drainage	10
VOLUME OF SEWAGE AND STORM WATER:—Water for domestic use eventually removed as sewage—Average consumption of water—Variable discharge of sewage—Maximum hourly sewage discharge—Estimated maximum rainfall to be removed from absorbent and non-absorbent surfaces—Maximum hourly volume to be carried by each drain	12
GRADIENTS OF DRAINS:—Their dependency on the contour of the ground and available fall—Drains to be self-cleansing—Compensation for increased friction—Additional allowance in inspection chambers—Allowance for intercepting traps—Adjustment of gradients under unfavourable circumstances—Flat gradients to be avoided—Maximum gradient for stoneware drains—Cast-iron pipes to be used for steep gradients	13

CHAPTER IV.

VELOCITY AND FLOW OF SEWAGE. SIZE AND DISCHARGE OF DRAINS.

VELOCITY AND FLOW OF SEWAGE:—Hydraulics, or the theory of the flow of water in relation to the flow of sewage—Mean velocity—Greatest and least velocities—Computation of the mean velocity—Velocity of flow dependent on gradients of drains—Effects of friction—Definition of hydraulic mean depth—Eytelwein's formula—Velocities necessary to remove different substances—Velocity required for self-cleansing drains—Table of gradients—Mnemonic notes 16

SIZE AND DISCHARGE OF DRAINS:—Drains to be as small as practicable—Computation of sizes of drains—Table of velocity and discharge of sewage from drains at different gradients—Important facts concerning the velocity and discharge of sewage—Determination of the size of drains—Minimum size for drains—Restrictions of local bye-laws .. 21

CHAPTER V.

THE VENTILATION, FLUSHING AND CLEANSING OF DRAINS.

THE VENTILATION OF DRAINS:—Importance of thorough ventilation—Difference of level between air inlets and outlets essential—Alternative arrangements respecting the direction of air currents—Method of ventilating storm-water drains—Ventilation of foul drains—Maximum distance to be allowed between air inlets and outlets—Provision of intermediate ventilating chambers in long drains—Different methods of arranging fresh air inlets—Remarks on fixing mica flap inlets—Size of air inlets and outlets—Precautions to be observed in fixing foul air outlets—Valves or cowls not desirable for drain ventilation 26

THE FLUSHING AND CLEANSING OF DRAINS:—Sanitary advantages of flushing—Various arrangements for flushing drains—Table showing volume of water required for flushing purposes—Underground flushing tanks—Periods of flushing discharge 33

CHAPTER VI.

FORM, MATERIALS AND JOINTS OF DRAIN PIPES.

FORM AND MATERIALS OF DRAIN PIPES:—Pipes of circular section chiefly used—Characteristics of good stoneware pipes—Earthenware pipes not to be used—Test for porosity—

		PAGE
Pipes for foul drains to be specially selected—Thickness and weight of stoneware pipes—Specification for cast-iron pipes—The Dr. Angus Smith and Bower-Barff preservative processes—Weight and thickness of cast-iron pipes—Iron pipes more reliable than stoneware for foul drains	.. ..	38
JOINTS OF DRAIN PIPES:—A thoroughly sound joint necessary—Joints for iron pipes—Spigot and socket joint for stoneware pipes—The double seal joint—Hassall's patent joint ..	..	43

CHAPTER VII.

LAYING DRAINS.

Necessity of sight rails—Their construction—Use of boning-staff—How to fix sight rails—Foul drains to be laid on concrete bed—Packing sides of pipes with concrete—Filling in—Drains laid near buildings to be surrounded by concrete—Drains not to be laid under buildings—Precautions to be adopted where no alternative arrangement is possible—Arches to be turned over drains passing through walls—The use of taper pipes 47

CHAPTER VIII.

INSPECTION AND INTERCEPTING CHAMBERS.

INSPECTION CHAMBERS OR MANHOLES:—Method of arranging the junctions of drains—The object of inspection chambers—Their size and construction—Arrangement of branch channels—Formation of benchings—Manhole covers—Inspection chambers formed with rock concrete tubes 54

INTERCEPTING CHAMBERS:—Disconnection of drains from public sewer—Construction of intercepting chambers—A self-cleansing trap necessary—Remarks respecting the cleaning arm of the intercepting trap 61

CONTENTS. xi

CHAPTER IX.

SOIL, VENTILATING AND WASTE PIPES.

	PAGE
SOIL PIPES:—Materials generally used—Lead soil pipes—Table of weights—Objections to lead soil pipes—Joints of lead pipes—Joint between lead and cast-iron pipes—Joint between lead and stoneware pipes—Advantages of cast-iron soil pipes—Table of weights—Preservatives—Joints—Cast-iron pipes to be fixed with holderbats or blocking pieces	62
VENTILATING PIPES:—Materials used — Precautions to be observed in the case of cast-iron pipes—Galvanised wrought-iron pipes	67
WASTE PIPES:—Air disconnection essential — Wastes to be trapped—Materials and sizes used for vertical wastes—Table of weights and thickness of drawn lead pipes—Sizes of lead waste pipes	68

CHAPTER X.

MANHOLE COVERS.

Form and size of covers—Direct ventilating cover—Solid manhole cover—Substances used for sealing the joint between cover and frame—Solid cover with sunk top—Advantages of sunk top—Double manhole cover—Double manhole cover with sunk top 71

CHAPTER XI.

TRAPS.

INTERCEPTING TRAPS:—Best form of trap or seal—Remarks respecting traps with water seal—The essentials of a good intercepting trap — Desirability of a cascade or waterfall action—Traps rendered more self-cleansing when formed

with a contracted throat—Circumstances under which traps without cascade action may be used	75
PLUMBERS' TRAPS :—Internal fitments to be trapped—D-traps objectionable—Defects of bell traps—Siphon traps—Anti-D traps—Water seal of traps broken by evaporation, momentum and siphonage—Arrangement and size of anti-siphonage pipes	78

CHAPTER XII.

SURFACE GULLIES. KITCHEN AND SCULLERY SINKS.

SURFACE GULLIES :—Description of gully to vary according to its situation—Storm-water gullies to be trapless—Rain-water shoes as an alternative—Necessary precautions in the use of trapless gullies—Gullies for foul drains to be trapped—Arrangement of gully with cleaning arm to branch drain—Difficulty of removing greasy liquids—Construction of grease traps—Objections to grease traps—Advantages of a flushing-rim grease gully with flushing cistern	84
KITCHEN AND SCULLERY SINKS :— General arrangement and construction — Anti-siphonage pipe to be provided—Sinks receiving greasy liquids should discharge into flushing-rim gullies	92

CHAPTER XIII.

WATER-CLOSETS.

Water-closets to be trapped and the soil pipe ventilated—The essentials of a satisfactory closet fitment—Objections to pan closets—Construction of plunger closets—Valve closets—Conditions governing the construction of a good valve closet—Defects of long and short hopper closets—Objections to wash-out closets—Merits of a well-designed wash-down closet—Constructional details relating to wash-down closets—Siphonic-action closets	94

CHAPTER XIV.

FLUSHING AND WATER-WASTE PREVENTING CISTERNS.

Water for drainage purposes to be distinct from domestic supply —Object of water-waste preventing cisterns—Quantity of water necessary for closet flushing purposes—The requirements of a good water-waste preventer—Valve closets to be provided with an after-flush water-waste preventer—Construction of a supply valve and bellows regulator—Sizes of supply valves and pipes for valve closets—Flushing cisterns for urinals 107

CHAPTER XV.

WATER-CLOSET CONNECTIONS.

Importance of proper connection between closet and soil pipe— Joint for lead pipe—Connection of lead to iron soil pipes, or lead to stoneware pipes—Method of jointing earthenware with lead, iron, or stoneware pipes—Connection of iron to stoneware soil pipes—Joint for iron to iron pipes—Joints of anti-siphonage pipes—Special form of joint made with an iron collar—Soldering lead pipes directly to earthenware— Connection between closet and soil pipe to be accessible— Description of joint to be avoided—Outlets of closets to be of P form 113

CHAPTER XVI.

BATHS.

Their hygienic importance—Descriptions of baths—Essentials of a sanitary form of bath—Slate baths—Porcelain baths— Marble baths—Zinc baths—Enamelled iron baths—Copper baths—Average size and weight of plunge baths—Objections to concealed standing overflow and waste—Good arrangement for bath overflow and waste—Alternative arrangement for bath overflow—Water supply to discharge *above* the overflow level of bath—Baths to be unenclosed—Cradling for baths 119

CHAPTER XVII.

HOUSEMAIDS' SLOP SINKS.

	PAGE
Combined closet and slop sink—Housemaids' washing-up sink—Combined slop closet and wash-up sink—Size of soil pipes for slop closets—Slop sinks to be in well lighted and ventilated lobbies	131

CHAPTER XVIII.

LAVATORIES.

Requirements of a good lavatory basin—Fixed basins with plain rim—"Flushing-rim" basins—"Tip-up" basins—Lavatory tops—Defects of lavatory basins with concealed overflow—Objections to basins fitted with concealed standing overflow and waste—Simple form of standing waste and overflow—Arrangement of lavatory fittings when a standing waste and overflow is not permitted—Water supply to lavatory basins—Lavatory wastes—Anti-siphonage pipe to be provided .. 135

CHAPTER XIX.

URINALS.

Composition of urine—Urinals generally unfavourably situated—Necessity for ample water supply—Objections to the use of closets as urinals—Essentials of a satisfactory urinal—Advantages of basin urinals for private residences—Good general arrangement of basin urinal—Objections to treadle-action flushing appliances—Urinal compartments to be constructed of non-porous materials—Enamelled iron urinal basins—Self-flushing urinal basins—Details respecting urinal divisions—Stall urinals—Stall urinals with semi-circular backs—Trough urinals—Automatic flushing tanks for urinals 143

CHAPTER XX.

TROUGH CLOSETS AND LATRINES.

PAGE

To be detached from inhabited buildings and well ventilated—Trough closets—Latrines—Latrine pans with flushing rims—Abundant water supply necessary—Latrines to be unenclosed—General construction of latrines—Disadvantages of trough closets and latrines—Arrangement of flushing apparatus   155

CHAPTER XXI.

TESTING NEW DRAINS.

The hydrostatic or water test—Method of application—The "sweating" of drain pipes—Tested pipes recommended to be used—Testing long drains—Testing drains when embedded in concrete—Good form of drain plug—The smoke test—Description of smoke machine—Points to be observed when testing with smoke   161

CHAPTER XXII.

EXAMINING AND TESTING OLD DRAINS AND SANITARY APPLIANCES.

Drains and fittings to be periodically examined and tested—Drains frequently defective from unforeseen causes—Importance of systematic and thorough examination—Method of conducting the examination of the external drainage system—Wells and underground tanks to be noticed—The hydrostatic and smoke tests—Testing old drains with smoke—The examination of internal sanitary fittings—The arrangement of the water supply to be ascertained—Smoke rockets—The scent test—Testing with oil of peppermint   168

INDEX   179

ILLUSTRATIONS.

FIG.		PAGE
1.	Drainage plan for small residence .. .. *Frontispiece*.	
2-5.	Intercepting chamber with small ventilating chamber at side	30
6-9.	Intercepting chamber with mica-flap fresh air inlet ..	31
10.	Automatic flushing chamber, fitted with Mr. Rogers Field's patent siphon and trapping box	36
11.	Section through joint for cast-iron pipes..	43
12.	Ordinary spigot and socket joint for stoneware pipes ..	44
13.	The "double-seal" joint for stoneware pipes	45
14.	"Hassall's patent double-lined joint" for stoneware pipes	45
15.	Section showing method of fixing sight rails, &c., for laying drains	48
16.	Transverse section through concrete bed for drain pipes	51
17.	Method of laying a taper pipe for drains..	52
18-21.	Inspection chamber or manhole	56
22.	Section of channel bend for sharp curves	57
23-26.	Inspection chamber or manhole for drains at right angles	59
27-30.	Inspection chamber formed with rock concrete tubes ..	60
31.	Section showing connection between lead soil pipe and cast-iron drain pipe	65
32.	Ventilating manhole cover with dirt box under	72
33.	Solid manhole cover	72

FIG.		PAGE
34.	Solid manhole cover with sunk top	73
35.	Double manhole cover..	73
36.	Double manhole cover with sunk top	73
37.	Intercepting trap with cascade or waterfall inlet	77
38.	Intercepting trap without cascade or waterfall inlet ..	78
39.	Sketch of obsolete D trap	79
40.	Section through bell trap	79
41–42.	Section through siphon traps fitted with screw cap ..	80
43.	Section through "anti-D" trap	80
44.	Method of fixing valve closets	81
45.	Sketch of closets on ground and first floor, showing method of fixing anti-siphonage pipe	82
46.	"Trapless" gully to receive rain-water only	85
47.	Section through rain-water shoe	86
48.	Trapped gully with cleaning eye attached	87
49.	Ventilated grease trap for scullery sinks	89
50.	Flushing-rim grease gully	90
51.	Arrangement of flushing-rim grease gully for scullery sinks, with automatic flushing cistern attached ..	91
52.	General arrangement of "pan" closet	96
53.	"Plunger" or "plug" closet	97
54.	"Trapless" plug closet	98
55–56.	"Long" and "short" hopper closets	101
57.	Section through "wash-out" closet	102
58.	Method of fixing "wash-down" pedestal closets	103
59.	General arrangement of "siphonic-action" closets.. ..	105
60.	Sketch of "supply valve and bellows regulator" for water supply to valve closets..	111
61.	Section showing connection between lead trap and iron soil pipe	115
62.	Connection between earthenware trap and iron soil pipe	116
63.	Section through submerged water-tight joint between earthenware trap and lead bend	116

LIST OF ILLUSTRATIONS. xix

FIG.		PAGE
64.	Sketch showing lead socket attached to earthenware closet by means of a solder joint	117
65.	Concealed standing overflow and waste to baths	124
66–66A.	Section and sketch showing visible standing overflow and waste to baths	126, 127
67.	Alternative arrangement for bath waste and overflow	128
68.	Sketch of bath fixed without any enclosure	129
69.	Arrangement of "cradling" for baths	130
70.	Sketch of slop top for water-closets	131
71.	Housemaids' washing-up sink with visible standing overflow and waste	132
72.	Combined housemaids' slop closet and wash-up sink	133
73.	Section through a "fixed" lavatory basin	136
74.	Lavatory basin with flushing rim	136
75.	Sketch of "tip-up" lavatory basin	137
76.	Lavatory basin with solid waste plug and inaccessible overflow	138
77.	Common arrangement of lavatory basin	139
78.	Lavatory basin with concealed standing overflow and waste	140
79.	Lavatory basin with standing waste and overflow	140
80.	Another form of lavatory basin with standing waste and overflow	141
81.	Sketch of flat-back lipped flushing-rim urinal basin	146
82.	Angular lipped flushing-rim urinal basin	146
83–84.	Elevation and section showing general arrangement of basin urinal suitable for a private residence	147, 148
85.	Sketch of base-plate for urinal	149
86.	Sketch showing a range of stall urinals	150
87.	A range of stall urinals with semicircular backs	151
88.	Section through stall urinal with semicircular back	152
89–90.	Sketch and section showing general arrangement of trough urinals	153, 154

FIG.		PAGE
91.	Sketch showing a range of trough closets	156
92.	Section through a range of latrines	157
93.	Transverse section through latrine pan	158
94.	Section through latrine pan with flushing rim	159
95.	Drain plug fitted with test cock	164
96.	Sketch of india-rubber drain bag with air pump	165
97.	Sketch of smoke-generating machine for testing drains	167

SANITARY HOUSE DRAINAGE.

CHAPTER I.

PRELIMINARY REMARKS. GENERAL PRINCIPLES.

PRELIMINARY REMARKS:—Necessity of a sanitary drainage system for the maintenance of health—Drains often laid with insufficient supervision—Attention to minor details an important matter—Explanation of the "conservancy" and "water-carriage" systems of drainage—The "water-carriage" system in general use.

GENERAL PRINCIPLES:—Definition of "sewage" and "storm-water" drainage—Object to be attained in every system of house drainage—Water supply to be distinct from drains—Drains, &c., to be placed outside buildings—Precautions to be adopted where this cannot be carried out—House drainage to be disconnected from public sewers—Purpose of disconnection—Thorough ventilation of drains necessary.

PRELIMINARY REMARKS.

THE intimate connection existing between the sanitary condition of the domestic drainage system and the physical well-being of those who reside immediately within its sphere of influence, is acknowledged by all who have given the subject serious attention. For this reason the provision of thoroughly efficient arrangements for the removal of sewage and waste matters constitutes one of the many important and necessary essentials for the maintenance of health. It is only within recent years that the principles and details connected with the construction of a really sanitary system for general domestic purposes have received that careful attention which the subject demands. Formerly, so long as the drains were out of sight, scarcely any thought was given to their fitness or condition for the proper fulfilment of their duties, until their insanitary condition was unpleasantly revealed by the presence of sickness or disease. Even now the drains of an ordinary dwelling are frequently laid with little or no supervision, and with scarcely any thought as to the best general arrangement that could be obtained.

It cannot be too strongly insisted upon that not only the general arrangement, but also all the minor details, of a drainage system shall be carried out in as perfect a manner as possible. A little extra attention or expense in the first construction will be amply repaid by the lessened risk of danger to health and inconvenience which is afterwards obtained. The selection of sanitary appliances which are in every way fitted for the particular purposes for which they are required, is oftentimes considered to be a matter of minor importance. For instance, the same form of closet apparatus is seen fixed in places where it will be subject to conditions which are totally different in each case. Whilst it may answer admirably in one instance, yet in the other it is probable that a different type of closet might have been adopted with advantage. It should therefore be remembered that no individual type of fitment—however good of its kind—is suitable for every situation and combination of circumstances which may arise in practice; and it is upon the exercise of sound judgment in the selection of suitable appliances that the efficiency of the internal sanitary arrangements of a dwelling will in a great measure depend.

The removal of domestic sewage matters is usually effected either by what is known as the "*conservancy system*" or the "*water-carriage system*."

The "conservancy system" embraces the dry earth, midden, cess-pit, and pail or tub methods of sewage removal. Such methods are often found convenient for small villages or isolated houses and institutions; in fact, they may provide the only practicable way of dealing with the sewage from such places. For large towns, however, there are serious objections to the use of the conservancy system, and the number of towns in which this system is adopted is rapidly diminishing, the water-carriage system of sewage removal being substituted.

In the water-carriage system the sewage matters and refuse liquids are deposited in conduits or pipes specially provided for the purpose, and are carried along by the velocity of the flow of water with which they are associated, or which is introduced into the drains for the purpose of so removing them. This system affords the most satisfactory method for the removal of sewage matters from dwellings, and as it is now in general use, the following observations will be confined to the consideration of that system so far as it relates to the sanitary requirements of domestic buildings.

General Principles.

The waste liquids or matters intended to be removed under ordinary circumstances may be broadly divided into two classes—viz. sewage or foul drainage, and rain or storm-water drainage. Under the term "sewage or foul drainage" is included impure liquids of every description, together with any fæcal and other refuse matters mixed with water—either in solution or suspension—that might affect the individual or general health. Storm-water drainage is confined to the conveyance of rain-water collected from roofs and other external surfaces of buildings, areas, yards, &c.

The primary object of any domestic drainage system is the safe and speedy removal of all waste matters to some convenient place where they may be collected for treatment without danger to health. For present purposes, this may be considered as having been accomplished when the waste matters have been properly delivered in the public sewer, the collection and disposal of them from that point being undertaken by the local authorities. To achieve this end, it is of highest importance that certain general principles shall be efficiently carried out, viz.:—

1. The drainage system must be entirely disconnected from the domestic water supply. Under no circum-

stances should the water supply become liable to contamination either by contact with sewage itself, or by the absorption of any gases that might arise therefrom.

2. The whole of the drains and their appurtenances must, where possible, be so arranged as to be *outside* all buildings. This, however, can only be complied with in a modified form, as it is found necessary for the sake of convenience, to place many of the fitments (as water-closets, sinks, baths, &c.) used in connection with a drainage system, within the building. These fitments should be so arranged as to be adjacent to an external wall. Housemaids' sinks, water-closets, &c., are also preferably placed in an annexe, with a well-ventilated lobby separating them from the main building. Where internal sanitary fitments are used, they must in every case be completely disconnected from the drains, either by an interval of external air, or by the passage of a continuous current of fresh air at its junction with the drain.

3. The domestic drainage system must be definitely cut off from the general drainage of the district, so as to prevent the passage of sewer air or foreign matters from the local sewers into the house drains.

4. Every provision must be made for the circulation of a continuous current of fresh air through all the drains comprised within the system.

The degree of perfection in which these general principles are carried out will be the measure of the sanitary efficacy of the whole.

CHAPTER II.

SYSTEM OF DRAINAGE. STORM-WATER SECTION. FOUL-DRAINAGE SECTION.

SYSTEM OF DRAINAGE:—Objections to storm-water gullies connected directly to foul drains—Foul and storm-water drains to be grouped into distinct sections—Disposal of storm-water—Disconnection of storm drains from foul drains—Distribution of drainage under the separate system.

STORM-WATER SECTION:—Rain-water only admissible—Gullies to be trapless — Intercepting chamber required at junction with foul drainage section—Free ventilation to be provided through all storm-water drains.

FOUL DRAINAGE SECTION:—All gullies to be trapped—Ventilation of drains—Precautions to be observed respecting internal fittings—Disconnection from public sewer.

SYSTEM OF DRAINAGE.

IT is a very general practice to connect surface gullies receiving storm-water only, direct to foul drains. This method cannot be too strongly condemned, for in dry weather the traps of these gullies become unsealed by evaporation, and permit the escape of impure air from the drains at points where such escape may be a possible source of danger to health. Another objection to the connection of storm-water gullies direct to foul drains, is that the branch drains by which they are connected have, in most cases, no current of air passing through them. Being trapped at the gully end, these branches act as reservoirs of stagnant sewer-air from the foul drains when they are not a source of danger by being unsealed in dry weather. It is, therefore, desirable that the foul and storm-water drains be grouped together, so that each may form a distinct section of the drainage system.

The foul-drainage section should convey the whole of the sewage and other impure matters, whilst the storm-water section receives only rain-water. Any storm-water areas or gullies receiving water that is liable to be fouled, such as surface gullies to areas or yards, must be connected with the foul-drainage section. Such gullies should be arranged, if practicable, to receive frequent discharges of waste water in addition to storm-water, so as to prevent them becoming unsealed during dry periods.

After collecting the various foul and storm-water drains into their respective sections, the storm-water may be conveyed to an underground tank and stored for use. In places where the rain-water is not required, the storm-water section will discharge into the foul drain at some convenient point, care being taken to provide adequate disconnection at the point of discharge by means of an intercepting chamber. The whole of the drainage, having thus been collected into one main drain, is discharged into the public sewer, but disconnected therefrom by means of another intercepting chamber. The Frontispiece (Fig. 1) illustrates the method of grouping the foul and storm-water drains into sections.

In localities where the *separate* system has been adopted for the general drainage of the district, the two sections would be kept distinct throughout, the storm-water section discharging into the storm sewers, and the foul-drainage section into the foul-drainage sewers.

The Storm-Water Section.

Nothing should be allowed to enter these drains except uncontaminated storm or rain-water. Any gullies or areas receiving rain-water which is liable to become fouled, must be connected to the foul-drainage, and not to the storm-water section. All gullies on the storm-water section should be *trapless*, but may be provided with a silt-pit, in which any sediment or heavy substances, such as sand, &c., may be retained for periodical removal. The intercepting chamber to the storm-water section at its junction with the foul drain must be provided with an open grating or ventilating chamber, so that a current of air may pass freely through all the drains comprised in this section.

The Foul-Drainage Section.

The whole of the gullies connected with the foul drains must be *trapped*. The branch drains should be as short as possible; but at the same time the collecting drain for the branches should be kept at least 8 feet away from the walls of a building, to avoid the risk of sewage soaking into the basement from a defective pipe or joint.

It is not often practicable, although theoretically desirable, to ventilate each small branch; but it is imperative that at least the head of each collecting drain shall be open, to allow a current of air to circulate through them. Where branch drains are more than 20 feet long they should be ventilated.

As far as possible, all sanitary fitments within the building should discharge in the open air over trapped gullies. Where this cannot be carried out, as in the case of water-closets, housemaids' sinks, &c., they should be connected with the *untrapped and ventilated* head of a drain or branch. The passage of any sewer air into the building through the fitment must also be further safeguarded by the proper construction of the fitment itself.

The foul drainage must be distinctly disconnected from the public sewer by means of an intercepting chamber, and provision also made at this point for the admission of a continuous current of fresh air for circulation through the drains.

CHAPTER III.

GENERAL PLAN. VOLUME OF SEWAGE AND STORM-WATER. GRADIENTS OF DRAINS.

GENERAL PLAN :—Its preparation—Arrangement of drains—Inspection chambers—Foul-air outlets—Stable drainage to be distinct from house drainage.

VOLUME OF SEWAGE AND STORM-WATER :—Water for domestic use eventually removed as sewage—Average consumption of water—Variable discharge of sewage—Maximum hourly sewage discharge—Estimated maximum rainfall to be removed from absorbent and non-absorbent surfaces—Maximum hourly volume to be carried by each drain.

GRADIENTS OF DRAINS :—Their dependency on the contour of the ground and available fall—Drains to be self-cleansing—Compensation for increased friction—Additional allowance in inspection chambers — Allowance for intercepting traps — Adjustment of gradients under unfavourable circumstances—Flat gradients to be avoided — Maximum gradient for stoneware drains — Cast-iron pipes to be used for steep gradients.

GENERAL PLAN.

A GENERAL site plan of all the buildings for which drainage is required should be prepared, and the position of every soil pipe, slop sink, waste, rain-water pipe, surface gully, &c., shown thereon.

The position of the junction with the public sewer having been fixed, a series of levels are required to ascertain the relative heights of the various gullies, &c.,

and the amount of fall available between them and the outfall.

Having decided which of the wastes, soil pipes and gullies shall be considered as discharging foul or storm-water respectively, they may be grouped together in the most convenient manner, so as to complete the foul and the storm-water sections. When it is not intended to store the rain-water in an underground tank, the foul and storm-water sections are brought together at some convenient point, and then carried direct to the proposed junction with the public sewer.

In determining the lines of drainage, due regard must be given to the natural contours of the ground, and the falls available for each of the branch drains. They should be laid in perfectly straight lines, with an even gradient from point to point. Wherever one drain joins another, or any change—either in direction or gradient—takes place, an inspection chamber should be provided, except, perhaps, in the case of an unimportant branch. This is required so that all the drains may be accessible for examination and cleansing at any future time. Where long lines of drains occur, a series of inspection chambers must be arranged about 100 feet apart to allow of every portion of the drain being reached by drain rods if necessary. The drains should be designed and constructed not only to allow a current of air to pass through them, but also to permit the escape of vitiated air at such points where its discharge will not in any way be offensive or injurious to health.

With regard to stable drainage, it is better to keep it entirely separate from the house drains, and arrange for an independent outfall to the public sewer.

Volume of Sewage and Storm-Water.

To determine the proper sizes for the various drains, it is necessary to calculate the maximum amount of drainage to be removed by each drain. Dealing first with sewage proper, it has been found that the whole of the water supplied for domestic purposes will practically be removed in some form or other as "sewage." The average daily consumption of water for domestic purposes is about 30 gallons per head. Provision must therefore be made for the removal of 30 gallons (or about 5 cubic feet) of sewage per head of occupation in every twenty-four hours. As the discharge of sewage is not regularly distributed over the twenty-four hours, but varies considerably at different periods of the day, it is usual to assume that half the total daily discharge will pass through the drains in six hours. This gives $2\frac{1}{2}$ gallons of sewage per hour for each occupant as the maximum hourly discharge that may require to be removed.

With regard to storm-water, for large areas of

partially absorbent surfaces, such as gravelled yards, &c., it is sufficient to consider $\frac{1}{2}$ inch per hour as the maximum amount of storm-water to be carried from such areas. For roofs, stone or concrete surfaces of yards, and other similar impervious surfaces, it is advisable to provide for a maximum rainfall of 1 inch per hour. In other words, provision must be made for removing half a gallon of storm-water per hour from each superficial foot in area of the roofs and yards served by the drain under consideration.

The maximum hourly volume of sewage and storm-water to be discharged at the outfall can thus be readily calculated, and also the maximum hourly volume to be carried by each branch drain comprised within the foul or storm-water sections.

Gradients of Drains.

The gradients that may be given to the drains will depend on the natural contours of the ground and the fall available between the outfall and the various points of discharge to be connected thereto. As far as possible, all drains should be laid with *self-cleansing* gradients—i.e. laid to such falls that the velocity of the sewage or storm-water flowing through them under normal conditions is such as to keep the drains free from any deposit.

When calculating the gradients that may be given to each drain, it must be borne in mind that wherever there is a change of direction additional fall must be given to compensate for the increased friction at such point, so that the velocity of the flowing sewage may not be reduced. For this purpose an additional fall of $1\frac{1}{2}$ inches at every junction or change of direction will generally be sufficient. Where the junction occurs within a manhole or inspection chamber, an extra allowance equal to half the diameter of the main drain must be provided, in order that the minor junction or branch may discharge *over* the main channel. At points where an intercepting trap is placed, an allowance of 3 inches must be made for the difference of level between the inlet and outlet of the trap.

Should local circumstances not admit of the whole of the drains being laid to self-cleansing falls, it is preferable to lay the main drain at a flat gradient than to sacrifice the self-cleansing gradients of the branches to admit of a slight additional fall being given to the main drain. In cases where it is found necessary to lay the main drain at a flat gradient, a flushing chamber should be provided at its head, so that it may be periodically flushed and cleansed.

When sufficient fall is otherwise obtainable, it is in every way a false economy to endeavour to save a few inches—or even feet—in the depth of excavations required for the drains, if it is effected at the expense of forming flat or insufficient gradients. Provided the drains are properly laid, with all necessary inspection chambers, &c., there should afterwards be no necessity

GRADIENTS OF DRAINS.

to disturb any portion of them. But if the gradients are insufficient, or the system badly designed and carried out, the drains will become a constant source of annoyance and expense. It is not advisable, however, to lay stoneware drains at a steeper gradient than 1 in 10, for if greater falls than that be adopted, the glazed surface of the invert of the drain is liable to become worn off by the friction of passing substances such as sand, &c. In places where such steep gradients are required, cast-iron pipes should be used.

CHAPTER IV.

VELOCITY AND FLOW OF SEWAGE. SIZE AND DISCHARGE OF DRAINS.

VELOCITY AND FLOW OF SEWAGE:—Hydraulics, or the theory of the flow of water in relation to the flow of sewage—Mean velocity—Greatest and least velocities—Computation of the mean velocity—Velocity of flow dependent on gradients of drains—Effects of friction — Definition of hydraulic mean depth — Eytelwein's formula—Velocities necessary to remove different substances—Velocity required for self-cleansing drains—Table of gradients—Mnemonic notes.

SIZE AND DISCHARGE OF DRAINS:—Drains to be as small as practicable — Computation of sizes of drains — Table of velocity and discharge of sewage from drains at different gradients—Important facts concerning the velocity and discharge of sewage—Determination of the size of drains—Minimum size for drains—Restrictions of local bye-laws.

VELOCITY AND FLOW OF SEWAGE.

THE velocity and flow of sewage is, for all practical purposes, similar to that of water flowing under the same conditions. As drain pipes of circular section are almost universally used, it will be sufficient to consider the theory of the flow of water or sewage through circular pipes only.

When referring to the velocity or flow of sewage, it must be understood that the *mean* velocity of the fluid is always implied, as the flowing particles have a velocity varying at different points of the same cross section. This varying velocity is due to the friction of the fluid against the sides of the pipe or channel. The *actual* velocity of the flowing liquid is least at the points of contact with the pipe and greatest at the centre of the flow. The *mean* velocity of any liquid stream is found by dividing the volume of discharge by the area of the cross section of the stream, and is usually expressed in feet per second or feet per minute.

It will be readily seen that the velocity of any given depth of flow will vary *directly* as the inclination or gradient of the pipe, so that the greater the inclination the greater will be the velocity of the flow.

It should also be noted that, in consequence of the friction of the fluid particles against the sides of the pipe, the velocity of the flowing liquid through a pipe laid at any specific gradient varies according to the depth of flow on the invert of the pipe. For instance, in a pipe 6 inches in diameter it will be found that the velocity of a stream having a depth of 1 inch on the invert, is less than that of a stream having a depth of 2 inches on the invert. This is accounted for by the fact that the frictional surface of the pipe (or, as it is usually called, the wetted perimeter of the pipe) bears a larger ratio to the sectional area of the stream when its depth is only 1 inch on the invert than when the depth of the flow is 2 inches. This relationship between the wetted perimeter of a stream to its sectional area forms the basis of what is known as the "hydraulic mean depth" (usually written H.M.D.) of a stream. The hydraulic mean depth is found by dividing the sectional area of the stream by its wetted perimeter.

$$\text{H.M.D.} = \frac{\text{sectional area of flow}}{\text{wetted perimeter of flow}}.$$

In order to find the velocity of the flow of liquids through a circular pipe for any given gradient, numerous formulæ have been compiled from the results of various experiments. Amongst the best known may be mentioned those of Kutter, D'Arcy, Prony, Neville, Weisbach, Hawksley, and Eytelwein. Some of these are of a very complicated character, necessitating a long series of calculations. That of Weisbach, which is generally recognised as giving the most accurate results of all the formulæ mentioned, includes a distinct coefficient for friction for every change of velocity.

For present purposes, Eytelwein's formula has been selected as the most suitable, on account of its well-known character and simplicity of application—viz.:

$$V = 55 \sqrt{H \times 2 F}.$$

Where V = velocity in feet per minute;
 H = hydraulic mean depth in feet;
 F = fall in feet per mile.

The drains should be arranged with such gradients that any solid substances usually found in connection with domestic sewage, such as sand, pebbles, paper, fæcal deposits, &c., may be carried through them by means of the velocity or power of the flow of the liquids accompanying them.

From the results of various experiments carried out in ordinary well-constructed drains, it has been found that water flowing with a velocity of 120 feet per minute will overcome the resistance offered by coarse ballast or rounded pebbles, and remove them from the drain. When flowing with a velocity of 180 feet per minute, the liquid will remove small stones, fæcal matter, paper, or other substances of a like nature. The following table shows the velocity of the flow of water required to remove different substances:—

TABLE SHOWING THE VELOCITIES NECESSARY TO REMOVE DIFFERENT SUBSTANCES FROM PIPE DRAINS.

Description.	Velocity of Flow of Water.
Mud, liquid earth, &c.	15 ft. per minute
Clay	30 ,, ,,
River sand, grit, or small gravel	60 ,, ,,
Coarse ballast	120 ,, ,,
Sea shingle about 1 inch diam.	130 ,, ,,
Large shingle	180 ,, ,,
Angular flints, the size of a hen's egg	200 ,, ,,
Broken stones	240 ,, ,,

From an examination of the foregoing table it may be assumed that a velocity of flow which will remove large shingle will be quite sufficient to remove fæcal deposits and other solids that may require to be dealt with in an ordinary drain. Accordingly, all drains should, as far as possible, be laid with a fall sufficient to ensure a velocity of 180 feet per minute when the normal quantity of sewage is passing.

For general purposes, the normal quantity of sewage ordinarily passing through domestic drains is taken to be equivalent to a stream of sewage having a depth of one-quarter the diameter of the pipe through which it is flowing.

In order that drains may be self-cleansing under these circumstances, it is necessary that a stream of sewage having a depth of one-fourth the diameter of the drain must flow with a velocity of 180 feet per minute, so as to remove the solids with which it is usually associated. The minimum gradients that can be given for self-cleansing drains are as follows:—

TABLE OF GRADIENTS NECESSARY TO PROVIDE SELF-CLEANSING DRAINS WHEN THE DEPTH OF THE FLOW OF SEWAGE IS ONE-QUARTER THE DIAMETER OF THE PIPE THROUGH WHICH IT IS PASSING.

Diameter of Drain.	Gradient of Drain.
4 inches	1 in 40
6 ,,	1 in 70
9 ,,	1 in 100

As an aid to memory, the so-called decimal rule for self-cleansing gradients of drains is easily remembered, and well adapted for ordinary use, viz. multiply the diameter of the pipe by 10, and the result will determine the gradient. Thus:—

Diameter of Drain.	Gradient of Drain.
4 inches	1 in 40
6 „	1 in 60
9 „	1 in 90

On comparison with the table previously given, it is seen that this rule will afford good self-cleansing gradients to the drains.

Size and Discharge of Drains.

Where insufficient care is exercised in the determination of the sizes of drains required for any particular purpose, there is a tendency to make them much larger than is absolutely necessary for the work they may be called upon to perform. This is a great mistake, both on sanitary and economic grounds. The efficiency of a drain is not increased by its being larger than absolutely required; on the contrary, it is greatly impaired. For example, the normal quantity of sewage passing through a 4-inch drain may be quite sufficient to allow of its being self-cleansing, whereas the same quantity of sewage passing through a 6-inch drain laid at the same gradient would probably have insufficient depth on the invert to secure the necessary self-cleansing velocity. The larger drain would therefore remain in a more or less foul condition, whilst the smaller drain, in the same situation, would be comparatively clean. In addition to this, the first cost of the smaller and more efficient drain would be much less.

The gradients having been determined according to the fall available, together with the maximum volume of sewage required to be taken by each drain, the necessary sizes of the drains may be calculated by the formula—

$$D = V \times A.$$

Where D = discharge in cubic feet per minute;
V = velocity in feet per minute;
A = sectional area of flow in feet; or
Discharge in gallons per minute = $6 \cdot 25 \times V \times A$.

For the sake of convenience and ready reference, the following table has been prepared, showing the velocity and discharge of drains with various gradients and for different depths of flow. These have been calculated from the formulæ already given.

By a careful examination of the following table several important facts relating to the velocity and discharge of liquids through circular pipes will be

TABLE OF VELOCITY AND DISCHARGE OF SEWAGE FROM CIRCULAR DRAIN PIPES FOR VARIOUS GRADIENTS AND DEPTHS OF FLOW.

Diameter of drain pipe.	Depth of flow on invert of pipe.	Area of flow.	Hydraulic mean depth.	Fall 1 in 40.		Fall 1 in 50.		Fall 1 in 60.		Fall 1 in 70.		Fall 1 in 80.		Fall 1 in 90.		Fall 1 in 100.		Fall 1 in 110.	
				Velocity per Minute.	Discharge per Minute.	Velocity per Minute.	Discharge per Minute.	Velocity per Minute.	Discharge per Minute.	Velocity per Minute.	Discharge per Minute.	Velocity per Minute.	Discharge per Minute.	Velocity per Minute.	Discharge per Minute.	Velocity per Minute.	Discharge per Minute.	Velocity per Minute.	Discharge per Minute.
		ft.	ft.	ft.	gal.	ft.	gal.	ft.	gal.	ft.	gal.	ft.	gal.	ft.	gal.	ft.	gal.	ft.	gal.
4 in.	1 in. or $\frac{1}{4}$ dia. of pipe	·017	·049	197	21	177	19	161	17	..	..	..	..	..	..	..	..	..	..
	2 in. or $\frac{1}{2}$,,	·044	·083	257	71	230	63	210	58	..	..	..	..	..	..	..	..	..	..
	$3\frac{1}{3}$ in. or $\frac{5}{6}$,,	·078	·101	284	138	254	124	232	113	..	..	..	..	..	..	..	..	..	..
	$3\frac{1}{2}$ in. or $\frac{7}{8}$,,	·081	·100	282	143	252	128	230	117	..	..	..	..	..	..	..	..	..	..
	$3\frac{2}{3}$ in. or $\frac{11}{12}$,,	·084	·098	279	146	250	131	228	120	..	..	..	..	..	..	..	..	..	..
	4 in. or flowing full	·088	·083	257	142	230	126	210	116	..	..	..	..	..	..	..	..	..	..
6 in.	$1\frac{1}{2}$ in. or $\frac{1}{4}$ dia. of pipe	·038	·073	..	..	..	..	197	47	182	43	170	40	..	..	..	..	..	..
	3 in. or $\frac{1}{2}$,,	·098	·125	..	..	..	..	258	158	238	146	223	137	..	..	..	..	..	..
	5 in. or $\frac{5}{6}$,,	·175	·155	..	..	..	..	287	314	265	290	249	272	..	..	..	..	..	..
	$5\frac{1}{4}$ in. or $\frac{7}{8}$,,	·182	·151	..	..	..	..	283	322	262	298	245	279	..	..	..	..	..	..
	$5\frac{1}{2}$ in. or $\frac{11}{12}$,,	·188	·147	..	..	..	..	279	328	258	303	242	284	..	..	..	..	..	..
	6 in. or flowing full	·196	·125	..	..	..	..	258	316	238	292	223	274	..	..	..	..	..	..
9 in.	$2\frac{1}{4}$ in. or $\frac{1}{4}$ dia. of pipe	·086	·109	..	..	..	..	..	..	..	..	..	..	196	105	186	100	178	96
	$4\frac{1}{2}$ in. or $\frac{1}{2}$,,	·221	·187	..	..	..	..	..	..	..	..	..	..	257	355	244	337	233	322
	$7\frac{1}{2}$ in. or $\frac{5}{6}$,,	·394	·228	..	..	..	..	..	..	..	..	..	..	284	699	270	665	257	633
	$7\frac{7}{8}$ in. or $\frac{7}{8}$,,	·410	·225	..	..	..	..	..	..	..	..	..	..	282	723	268	687	255	653
	$8\frac{1}{4}$ in. or $\frac{11}{12}$,,	·424	·221	..	..	..	..	..	..	..	..	..	..	280	742	266	705	253	670
	9 in. or flowing full	·442	·187	..	..	..	..	..	..	..	..	..	..	257	710	244	674	233	644

observed, amongst which may be mentioned the following, viz. :—

1. The H.M.D. of a circular pipe flowing full is exactly the same as when flowing half-full—i.e. one-fourth the diameter of the pipe; consequently the mean velocity of the flow in both cases will be found to be the same.

2. The H.M.D. for liquids flowing through a circular pipe is *greatest* when the depth of the flow is approximately *five-sixths* of the diameter of the pipe, and it is at this point that the *maximum mean velocity* of flow is obtained.

3. The *maximum* discharge from a circular pipe is obtained when the depth of the flow is about *eleven-twelfths* of the diameter of the pipe, and not when flowing full, as might be supposed. This loss of discharge when a pipe is flowing full is due to the increased friction offered by the wetted perimeter of the pipe, as compared with the sectional area of the flow.

4. The volume of discharge of a circular pipe flowing full is exactly double that when flowing half-full.

Taking into consideration the facts that the maximum velocity of flow is obtained when the depth of flow is five-sixths the diameter of the pipe, and that the maximum discharge is obtained when the depth of flow is eleven-twelfths the diameter of the pipe, it is desirable that the drains should be of such a size as to discharge the previously ascertained maximum volume of sewage or storm-water per minute when the depth of the flow is *seven-eighths* the diameter of the pipe. This, being the mean proportion between five-sixths and eleven-twelfths, will permit of a slight increase in the volume of discharge without the pipe flowing full, whilst at the same time any slight decrease in the volume of discharge will increase the velocity of the flow, and so increase the scour and self-cleansing action of the sewage passing through the pipes.

No drain should, however, be less than 4 inches in diameter. In all cases of ordinary domestic drainage—even of a large institution—where the drains are laid to self-cleansing falls, it will be found that a 4-inch drain is large enough for the branches, and also for most of the collecting drains of the foul and storm-water sections, with, perhaps, a 6-inch or 9-inch main drain to the outfall.

In some districts the bye-laws of the local sanitary authority insist upon all the soil drains being not less than 6 inches diameter. Of course, under such circumstances, the evils entailed by this regulation cannot be avoided.

CHAPTER V.

THE VENTILATION, FLUSHING AND CLEANSING OF DRAINS.

THE VENTILATION OF DRAINS:—Importance of thorough ventilation—Difference of level between air inlets and outlets essential—Alternative arrangements respecting the direction of air currents—Method of ventilating storm-water drains—Ventilation of foul drains—Maximum distance to be allowed between air inlets and outlets—Provision of intermediate ventilating chambers in long drains—Different methods of arranging fresh-air inlets—Remarks on fixing mica-flap inlets—Size of air inlets and outlets—Precautions to be observed in fixing foul-air outlets—Valves or cowls not desirable for drain ventilation.

THE FLUSHING AND CLEANSING OF DRAINS:—Sanitary advantages of flushing—Various arrangements for flushing drains—Table showing volume of water required for flushing purposes—Underground flushing tanks—Periods of flushing discharge.

THE VENTILATION OF DRAINS.

IT is important that all drains shall be open to the air at both ends, and so arranged that a current of fresh air may be continuously passing through them, so as to secure their sanitary efficiency.

Where this course is properly carried out, the sewage becomes partially oxygenated whilst passing through the drains, and any gaseous products are carried into the open air at some convenient point and innocuously disseminated. For the thorough ventilation of drains it is not sufficient that the drains shall be open at each end, but there must also be a considerable difference of level between the openings, in order that a current of air may be induced by the difference of level. For this reason the fresh-air inlets of the drains should be situated at a low level, whilst the vitiated-air outlets should be at the highest level obtainable.

It is desirable that the direction of the air currents within the drains may be the same as that of the flow of the sewage, so that when a quantity of sewage is discharged into the drain, the displaced air may be carried with it and escape at the extracting pipe. In addition to this, the flowing sewage tends to induce a current of air within the drains in the same direction as the flow.

It is seldom, however, that the ventilation of drains can be safely and effectively carried out in this manner. It may be that the head of the drain is the top of a water-closet soil-pipe, and it would probably be undesirable (on account of the adjacent windows, &c.) to constitute this a fresh-air inlet to the drain without first carrying the soil pipe above the eaves of the house. Under such circumstances, to consider the head of this soil pipe as the low-level fresh-air inlet would necessitate the erection of an extraction pipe of great height at the lower or discharging end of the drain, in order to obtain a difference of level between the inlet and outlet sufficiently satisfactory to ensure the production of a current of air within the drain. In some instances, however, the local circumstances admit of low-level fresh-air inlets being placed at the head of the drains, whilst a high-level extracting pipe or shaft may be conveniently fixed at the outfall.

Generally, it is necessary to provide the low-level fresh-air inlet for the drainage system at the intercepting chamber, and the high-level extracting pipes or shafts at the head of each drain.

The thorough ventilation of every drain within the storm-water section can be easily accomplished. A fresh-air inlet is provided at the storm-water intercepting chamber (see Frontispiece), whilst the head of each branch terminates with an *untrapped* surface gully or rain-water shoe.

In the case of the foul drainage section, it is not always practicable to arrange for every branch having through ventilation, but the number and also the length of such unventilated branches should be as small as possible. At the same time, the whole course of every collecting drain or branch drain of any importance should be thoroughly ventilated.

The fresh-air inlets and the vitiated-air outlets should be so arranged that the greatest distance between the inlet and its furthest outlet is not more than 300 feet. Where this distance is exceeded (as in a long length of drain), intermediate ventilating chambers or manholes should be provided in convenient situations, so that the length of drain ventilated by any given inlet or outlet may be kept within the limits mentioned.

In places where no inconvenience can result, the fresh-air inlet to the drain should be perfectly free and open. This is usually effected by covering the intercepting chamber with a perforated iron grating having a dirt-box under. When the intercepting chamber cannot be conveniently covered with a perforated cover, a small ventilating chamber may be constructed at the side of the intercepting chamber, and covered with an iron grating, as shown in Figs. 2 to 5.

If the fresh-air inlet is near a door, window, or placed in any other position where a back draught or reverse current would be undesirable, it must be so constructed as to prevent the egress of any air or gases from the drain, whilst at the same time allowing for an adequate admission of fresh air into the drain. The ventilating cover or chamber is omitted, and an inlet pipe carried from the intercepting chamber to an adjacent wall or other suitable position, and terminating with a mica flap valve, as shown in Figs. 6 to 9. The inlet pipe must enter the intercepting chamber as close to the surface of the ground as possible, so as to prevent any accumulation of sewer air in the space between the top of the pipe and the manhole cover. If heavy traffic is likely to pass over the site, it is necessary that the pipe should be about 2 feet below

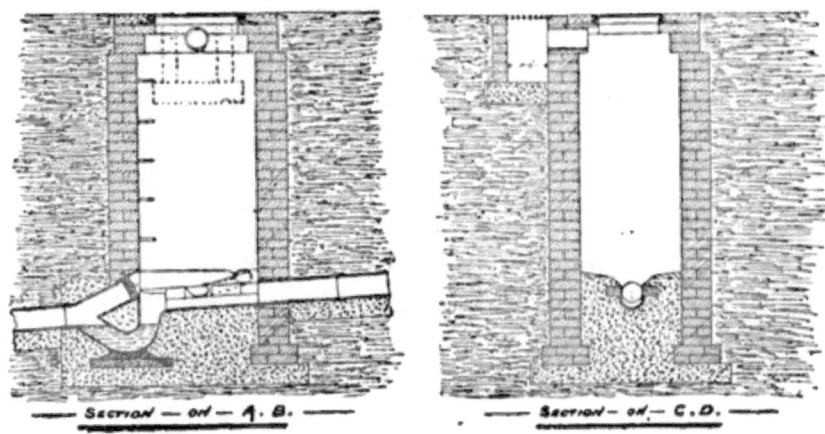

FIG. 2. FIG. 3.

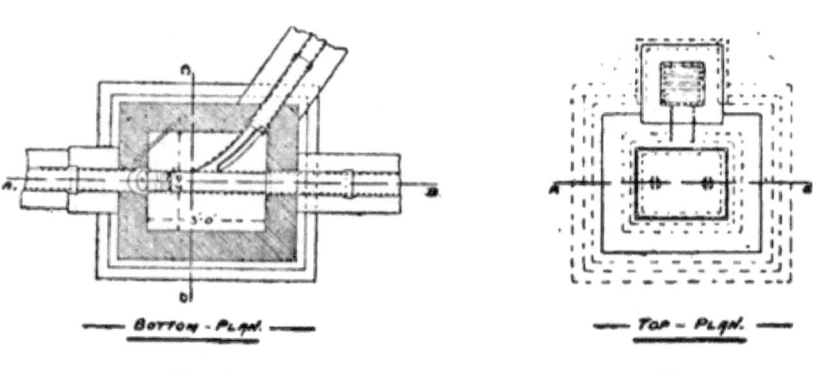

FIG. 4. FIG. 5.

THE VENTILATION OF DRAINS.

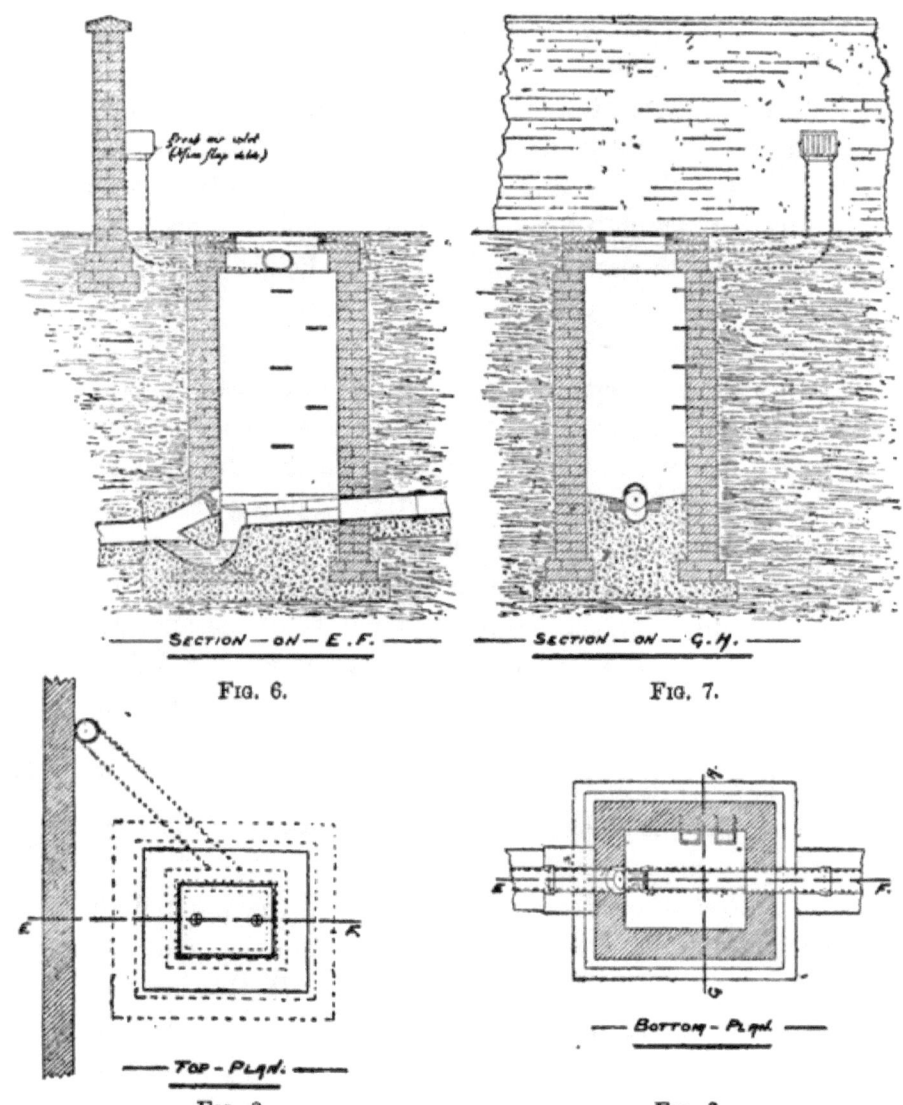

Fig. 6. Fig. 7.
Fig. 8. Fig. 9.

the surface of the ground to prevent it from being damaged or broken.

To assist the induction of a current of air, the mica flap inlet should be fixed in a comparatively cool place, and near the ground level, provided it is placed sufficiently high to be safe from any chance of being blocked with earth, rubbish, &c. A height of two or three feet from the ground will generally be found sufficient for this purpose.

It is a common practice to provide a fresh-air inlet much too small for the size of the drain it is intended to supply with air. For instance, a 4-inch diameter fresh-air inlet valve is sometimes provided to an intercepting chamber in which two 6-inch drains discharge, and both of which are dependent on the 4-inch inlet for the supply of air. The fresh-air inlet (whether it be a surface grating or mica flap valve) should in all cases have an effective sectional area of not less than the sectional area of the drain or drains discharging into the intercepting chamber. A 6-inch inlet pipe will suffice for one 6-inch drain, and a 9-inch inlet for two 6-inch drains.

The sum of the sectional areas of the vitiated-air outlets should be at least equal to the effective sectional area of the fresh-air inlets. If the head of each collecting drain or important branch has been carried up as an extracting shaft or vitiated-air outlet, it will generally be found that the combined sectional areas of these outlets will leave an ample margin.

The position and height of each extracting pipe should receive careful consideration, so that the air or gases escaping from the outlet may not prove a source of danger or annoyance. The pipes must be carried up perfectly straight for their whole height, no bends being permitted on account of any plinth, string-courses, cornices, eaves, &c., that may be met with.

The top of each extracting pipe should be not less than 6 feet above the eaves of the roof, or of any dormer window, with a minimum horizontal distance of 18 feet from any chimney or ventilator, so as to avoid the danger of down-draught. They should be so placed that the wind may blow freely across the top of the pipe, in order to induce an air current within the drains, and if the pipe can be carried higher than the ridge of the roof, so much the better.

As a broad principle, it is not desirable to provide any form of valve or cowl either to the inlet or outlet ventilator of the drains. In certain situations, however, (as already described), it is found necessary to fix a mica valve to the fresh-air inlet; but the top of the outlet pipe should be left quite open, and simply finished with a spherical copper-wire guard to prevent birds building nests therein.

The Flushing and Cleansing of Drains.

Although provision for flushing public sewers has for a long time received careful consideration, it is comparatively of recent date that the sanitary importance of thoroughly cleansing the house drains by means of a periodical flush has generally received sufficient attention. By means of the rapid and powerful discharge of a large volume of water, all deposits may be forcibly carried away, and at the same time the thorough renewal of the air within the drains greatly assisted.

A flushing tank or chamber should be provided at the head of every main or collecting foul drain, and it is desirable that this principle be carried out even though the drains are laid with ordinary self-cleansing falls. Arrangements should be made for a flushing tank to discharge over all gullies receiving greasy water, as from scullery sinks, &c., for the complete cleansing of the gully and branch drain connected thereto. In most cases it can generally be arranged that the branch drain from the scullery sink is also the head of one of the collecting foul drains, so that the flushing of this gully will suffice for flushing the collecting drain with which it is connected. (See Frontispiece.) Special flushing arrangements need not be made to unimportant branches, provided they are laid with self-cleansing falls and are of no great length. As a general rule, flushing tanks or chambers should be automatic in action.

The quantity of water required to properly flush a given drain will depend on its gradient and also on its

THE FLUSHING AND CLEANSING OF DRAINS.

length. The following table shows the amount considered necessary for the satisfactory flushing and cleansing of drains at different gradients up to a maximum length of 500 feet. It is seldom that any ordinary house drain will exceed that length.

TABLE OF VOLUME OF WATER REQUIRED FOR FLUSHING DRAINS UP TO 500 FEET IN LENGTH, WHEN LAID AT VARIOUS GRADIENTS.

Diameter of drain.	Gradient of drain.	Capacity of flushing tank or chamber.	Diameter of discharging outlet from flushing tank or chamber.
4 inches	1 in 40	30 gallons	3 inches
	1 in 50	40 ,,	3 ,,
6 ,,	1 in 60	60 ,,	4 ,,
	1 in 100	100 ,,	4½ ,,
	1 in 200	160 ,,	4½ ,,
9 ,,	1 in 100	200 ,,	6 ,,
	1 in 150	250 ,,	6 ,,
	1 in 200	300 ,,	6 ,,
	1 in 300	400 ,,	6 ,,

The method of flushing a gully receiving greasy water from a scullery sink will be further discussed when considering the form of gully required to receive scullery wastes.

In places where an iron flushing tank cannot be

conveniently fixed at the head of a drain, it becomes necessary to construct an underground flushing chamber. A section through an automatic flushing chamber fitted with Mr. Rogers Field's patent cast-

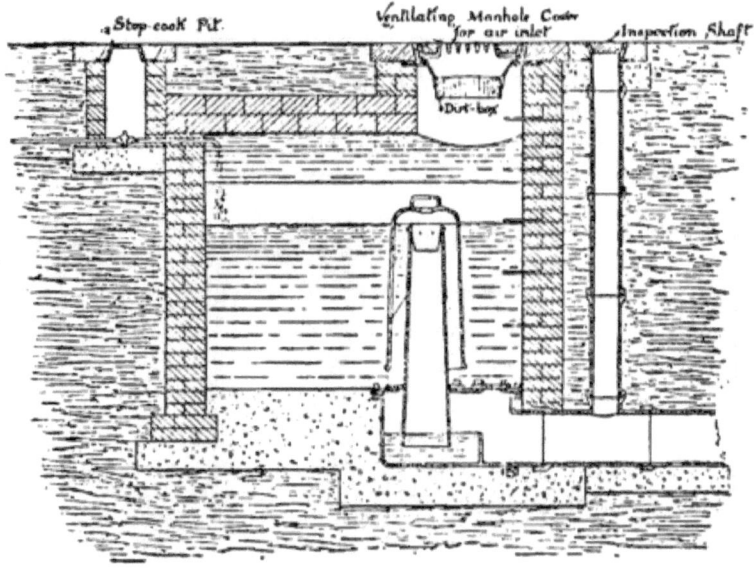

Fig. 10.

iron annular siphon and wrought-iron trapping-box is shown in Fig. 10.

When flushing tanks or cisterns are provided with covers—whether underground or not—care must be taken to allow the unrestricted entrance of air to the tank, or siphonic action cannot take place.

For small flushing tanks at the head of branch drains, it is generally sufficient to arrange for their discharge once a day. In the case of underground flushing cisterns at the head of main drains laid with self-cleansing falls, one discharge every five or six days may suffice.

CHAPTER VI.

FORM, MATERIALS AND JOINTS OF DRAIN PIPES.

FORM AND MATERIALS OF DRAIN PIPES:—Pipes of circular section chiefly used—Characteristics of good stoneware pipes—Earthenware pipes not to be used—Test for porosity—Pipes for foul drains to be specially selected—Thickness and weight of stoneware pipes—Specification for cast-iron pipes—The Dr. Angus Smith and Bower-Barff preservative processes—Weight and thickness of cast-iron pipes—Iron pipes more reliable than stoneware for foul drains.

JOINTS OF DRAIN PIPES:—A thoroughly sound joint necessary—Joints for iron pipes—Spigot and socket joint for stoneware pipes—The double seal joint—Hassell's patent joint.

FORM AND MATERIALS OF DRAIN PIPES.

FOR general purposes, drain pipes of a circular section have been found to give the best practical results, and are usually made either of cast iron or salt-glazed stoneware. Where drains and sewers of large size are required, they are constructed of brickwork and concrete of egg-shaped section, but drains of this description need not be considered in connection with ordinary house drainage.

Stoneware drain pipes should be salt-glazed, highly vitrified, impervious, perfectly smooth inside, true in section, straight in the barrel, with well-formed sockets, and an even thickness of material throughout. The pipes when struck should give a clear ringing sound, and when fractured present an almost metallic appearance. They should be entirely free from sand-holes, fire-cracks, or other defects.

Earthenware pipes should not be used, as they are very porous and are not burnt at a sufficiently high temperature to become vitrified. The degree of impermeability of a pipe may be ascertained by placing it in a vertical position, and temporarily blocking the lower end and filling the pipe with water. If the material is porous, and the pipe insufficiently glazed, the water will penetrate the pores of the material, and show itself on the outer surface of the pipe in the form of "sweat" or perspiration.

Where stoneware pipes are intended to be used for foul drains, they should be specially selected, and capable of withstanding a test of 25 feet head of water without showing signs of sweating. Most well-known makers manufacture a special class of pipe in which every length has been thoroughly examined and tested to a considerable head of water. Each pipe is distinctively stamped by the maker before being sent from the works. Stoneware pipes can now be obtained in 3-feet lengths. This is a great advantage and improvement, owing to the number of joints required being thus considerably reduced.

The following table shows the average thickness and weight of stoneware drain pipes:—

AVERAGE THICKNESS AND WEIGHT OF STONEWARE DRAIN PIPES.

Bore of pipe.	Net length of pipe when laid.	Length of socket.	Thickness of stoneware.	Average weight per pipe.
Inches.	Feet.	Inches.	Inch.	lb.
4	2	$1\frac{1}{2}$	$\frac{1}{2}$	18
6	2	$1\frac{3}{4}$	$\frac{5}{8}$	34
9	2	2	$\frac{3}{4}$	60

Cast-iron drain pipes should be in 9-feet lengths, of good tough grey iron from the second melting, smooth inside, true in section, perfectly straight in the barrel, with an even thickness of metal throughout, free from air-holes, sand-holes, and other defects. The sockets should be strong, with a good margin all round for caulking up, and the spigots provided with a bead cast on the end. They should be capable of withstanding 200 feet head of water, and coated with some preparation to prevent oxidation. The usual preservative processes employed are the "Dr. Angus Smith" and the "Bower-Barff."

To carry out the Dr. Angus Smith process, the pipes are carefully cleaned and scraped free from sand, scale, and rust. They are then dipped vertically into a mixture of pitch, coal-tar, and a small proportion of linseed oil. Whilst the pipes are being dipped the coating bath is maintained at a temperature of about 400° Fahr. After being allowed to remain in the bath about ten minutes, the pipes are withdrawn gradually, so as to allow the surplus mixture to run off.

When the process is properly carried out, the coating on the pipes should be tough and firmly adhering to the iron surfaces without any tendency to chip or scale off, having a uniform thickness throughout of $\frac{1}{100}$ inch. The pipes should be coated before leaving the foundry, so that the surfaces may not become oxidised or rusted before the coating is applied.

Where the Bower-Barff process is adopted, the pipes are thoroughly cleaned, placed in a chamber heated to 1200° Fahr. for eight or ten hours, and exposed to the action of superheated steam. By this means the surfaces of the pipes are completely covered with a hard coating of black oxide of iron, which in itself is stated to be totally unaffected by air or damp.

Another process recently introduced with much success, is to coat the interior of the pipes with a preparation of glass enamel, thus rendering them perfectly smooth for the passage of sewage, as well as preserving the interior surfaces of the iron. The exterior of the pipes should be well tarred with two coats of coal-tar before being laid.

The following table gives the weight and thickness of strong cast-iron pipes suitable for drain pipes :—

Table of Weight and Thickness of Cast-Iron Drain-Pipes.

Bore of pipe.	Net length when laid.	Thickness of metal.	Depth of socket.	Thickness of socket.	Weight per pipe.		
Inches.	Feet.	Inch.	Inches.	Inch.	Cwt.	qr.	lb.
4	9	$\frac{3}{8}$	3	$\frac{11}{16}$	1	1	20
6	9	$\frac{7}{16}$	$3\frac{1}{2}$	$\frac{13}{16}$	2	1	27
9	9	$\frac{9}{16}$	4	$\frac{13}{16}$	4	2	24

Glazed stoneware pipes are in general use for ordinary drainage purposes, and are well adapted for storm-water drains. For foul drains, however, cast-iron pipes are becoming increasingly used, on account of their clear bore, freedom from distortion, greater length, and consequently fewer joints. The joints of iron pipes, when well made, are perfectly air- and water-tight, and altogether more reliable than the joints of stoneware pipes.

The superiority of cast-iron drain pipes over stoneware has become generally recognised in the United States, and they are consequently employed to a very great extent, whilst in some districts the use of iron pipes is made compulsory by the regulations of the local authority.

Whether stoneware or iron drains be adopted for ordinary purposes, it is desirable that cast-iron drain pipes should invariably be used where the ground is soft, swampy, or treacherous, and in places where heavy traffic is anticipated, or where drains pass under buildings.

JOINTS OF DRAIN PIPES.

The efficiency of a pipe drain depends in a great measure on the character of the joint between the pipes. The whole of the joints when made must be perfectly air- and water-tight. The adjacent pipes should be concentrical with each other, and the bore of the drain quite even and smooth at the joints.

The joints of cast-iron drain pipes should be run with molten lead. To make the joint, a few gaskets of spun yarn are first driven in and well caulked. A band

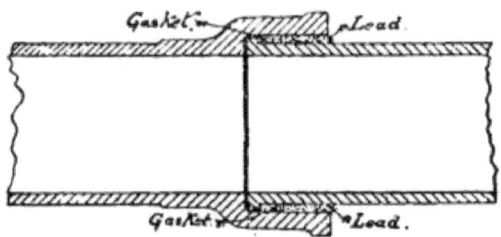

FIG. 11.

of well-tempered clay is then placed round the socket, leaving the interior clear to receive the lead. An opening is left in the clay at the top, through which the molten lead is poured. When the lead is cold the clay is removed and the joint caulked with a caulking tool. A section through the joint of a cast-iron drain pipe as ordinarily adopted is shown in Fig. 11. Sometimes a tight-fitting turned and bored joint is used to obtain better concentricity, the joint being then run with lead and caulked in the usual manner.

The jointing of stoneware pipes is a much more difficult matter to accomplish satisfactorily than in the case of cast-iron pipes. To overcome the difficulty numerous forms of joint and jointing material have been devised. Fig. 12 is a section of the common spigot and socket joint. The joint is generally made with cement only; but a better way is to force some

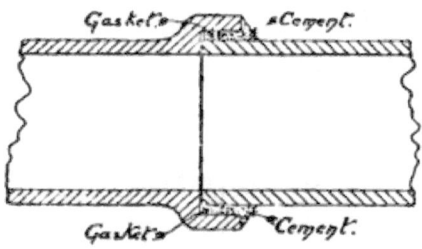

Fig. 12.

gaskets of yarn into the socket before jointing with cement. This prevents any portion of the cement protruding inside the pipe, and also more readily ensures the concentricity of the pipe.

The use of clay as a jointing material should not be allowed, owing to its solubility when in contact with running liquids. Neat Portland cement of good quality should be used, and it is advisable to confine the use of the ordinary spigot and socket joint to storm-water drains only.

JOINTS OF DRAIN PIPES.

A simple and good form of joint for stoneware pipes is shown at Fig. 13. It is called the "double seal" joint. A fillet of bituminous material (consisting of tar, sulphur, and ground pipe) is accurately cast on a

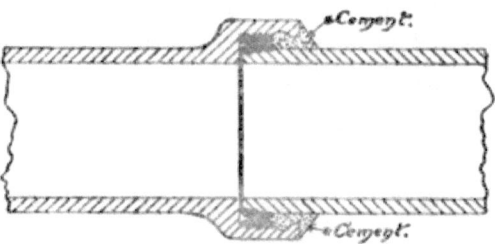

Fig. 13.

portion of each spigot and socket, so that when two pipes are pressed together, a perfectly water-tight and concentrical joint is obtained. Before jointing, the

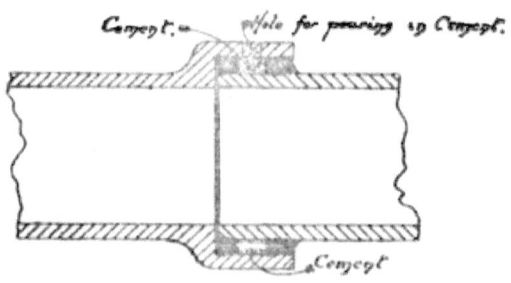

Fig. 14.

surface of the fillet must be smeared over with a mixture of 2 parts of tallow to 1 part of resin, which has been previously melted together. The remaining portion of the socket is filled with cement, the socket being undercut to afford a better hold for the cement. By this means the extra security given by two distinct jointing materials is obtained. A form of joint very similar to this is known as the " composite " joint.

"Hassell's patent double-lined joint" is another form that may be used with advantage (see Fig. 14). This consists of two fillets of bituminous material cast on each spigot and socket. When put together a central hollow space completely encircles the joint, which is filled with liquid cement poured through a hole provided at the top of the socket for that purpose. An impervious band of cement is thus obtained round the centre of each joint.

In addition to those mentioned, there are numerous other varieties of joint for stoneware pipes in the market; but space will not allow of any further reference to them.

CHAPTER VII.

LAYING DRAINS.

LAYING DRAINS:—Necessity of sight-rails—Their construction—Use of boning-staff—How to fix sight-rails—Foul drains to be laid on concrete bed—Packing sides of pipes with concrete—Filling in—Drains laid near buildings to be surrounded by concrete—Drains not to be laid under buildings—Precautions to be adopted where no alternative arrangement is possible—Arches to be turned over drains passing through walls—The use of taper pipes.

To ensure the pipes being laid straight and true in gradient, it is necessary to erect "sight-rails" along the line of the drain. Each sight-rail consists of two stout wooden uprights having a horizontal wooden rail well secured to them. The uprights are firmly planted in the ground at a sufficient distance on each side of the trench to be unaffected in any way by the excavations. The horizontal rail must be perfectly straight on its upper edge, and of sufficient depth and substance to avoid "sagging" when fixed. A "boning-staff" of suitable length must also be provided for use with the sight-rails. This is made in the shape of an ordinary T-square having a stout wooden shaft with a cross-head attached. The boning-staff is used when the pipes are being laid, so that the invert of the drain may be parallel to the line of sight between the sight-rails.

Care must be taken to fix the sight-rails at their proper relative levels. Having ascertained the reduced levels of the surface of the ground, and also those of the invert of the drain with respect to the Ordnance datum at the points immediately below the position of the proposed sight-rails, the sight-rails should be so fixed that the difference of level between the two rails is exactly the same as the difference of level required between the invert of the drain at the points immediately below the sight-rails. In other words, the sight-rails are fixed at such relative levels that the "line of sight" between them is parallel to the invert of the drain. The boning-staff must be of the same length as the distance between the sight-rail and the invert of the pipe immediately below it.

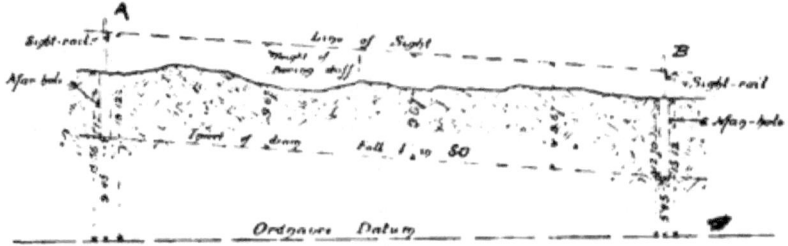

Fig. 15.

Fig. 15 shows the sight-rails fixed ready for use. In this instance the reduced levels of the invert of the drain, and the ground level immediately below the sight-rail A, are assumed to be 9·45 feet and 15·56 feet respectively. The levels below the sight-rail B are 5·45 feet and 12·70 feet. The rail A is proposed to be fixed 19·12 feet above datum. The invert of the drain having a difference of level between the sight-rails of 4 feet (9·45 − 5·45), it will consequently be necessary to fix the rail B at a height of 15·12 feet above datum, in order that the line of sight may be parallel to the invert of the drain. The length of the boning-staff for use in this case will be 9·67 feet, and the level of the invert of the drain at any point between the sight-rails can be readily ascertained by holding the boning-staff vertically at that point. When the head of the boning-staff coincides with the line of sight between the sight-rails, then the exact level of the invert of the drain will be indicated by the foot of the boning staff.

The trench should be excavated in a perfectly straight line to the necessary depth, the bottom being well rammed and levelled to the required gradient. It is desirable to lay all foul-water drains on a bed of concrete. Where the ground is ordinarily firm and solid, the thickness of the concrete bed need only be about 4 inches; but in situations where the ground is of a loose, wet or marshy description, then the concrete bed should have a minimum thickness of 6 inches. For storm-water drains the concrete bed may be omitted where the ground is fairly sound and firm.

When drains are being laid without a concrete bed, care should be taken that the trenches are not made deeper than absolutely required. Where the excavations have been made too deep, and the bottom of the trench requires to be made up in places, concrete should be used, for if made up with rammed earth, settlements of the drain pipes frequently occur at such points after the trenches are filled in and the ground has become consolidated.

The bottom of the trench should be well rammed to receive the concrete bed. The concrete should be 12 inches wider than the bore of the pipe, and the surface accurately brought up to the required gradient. Wooden moulds about 5 inches wide, 2 inches deep, and of a length equal to the width of the concrete, should be inserted in the concrete bed, and spaced 2 feet apart, centre to centre. These moulds are removed when the pipes are being laid, so that a depression is formed in the concrete sufficiently wide and deep to admit of the pipes being readily inserted and jointed on the under side, whilst the barrel of the pipe rests solidly on the concrete bed itself.

After the whole length of drain has been laid and tested, the sides of the pipes should be haunched up with concrete to a height of half the diameter of the pipe, as shown in the following sketch (Fig. 16).

This additional packing of concrete at the sides greatly supports and strengthens the drain, and at the same time forms an additional safeguard to the joints of the pipe.

Where no concrete bed is provided, slight sinkings

should be cut across the trench to allow for the sockets, so that the barrels of the pipes may lie solid on the bottom of the trench. In all cases care should be taken that the socket of each pipe is laid towards the head of the drain.

When the pipes have been satisfactorily laid and tested, the earth should be very carefully filled in over the pipes at first, and gently battened down with the spade. All large stones should be removed from the preliminary earth filling, and ramming should not be

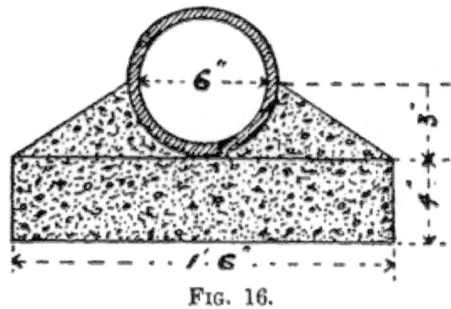

Fig. 16.

allowed until at least 2 feet of earth has been placed over the pipes.

Should it be necessary to lay a drain near the walls of a building (more especially if the building contains a basement), the drain must be entirely encased with concrete not less than 6 inches thick, so that no sewage from any defective joints or pipes may percolate through the foundations or walls to the interior of the building.

Unless absolutely unavoidable, drains should never

pass under a building. If, however, there be no other alternative, then the drain should be constructed with stout cast-iron pipes, coated with some preservative, and the joints run with lead and caulked. The pipes should be laid within a well-constructed culvert, so arranged as to be open to the air at both ends. This can usually be effected by forming a proper ventilated inspection chamber at each end of the culvert. Instead of the pipes resting on the bottom of the culvert, they may be laid on tee-iron bearers built into the sides of

FIG. 17.

the brickwork, so as to raise them about 9 inches from the bottom. Every portion of the drain is then accessible throughout its length, and sufficient space left all round to allow of the pipes and joints being thoroughly examined at any time. Where the drain passes under an unimportant or detached outbuilding, it will be sufficient to surround the pipe with 6 inches of concrete.

Relieving arches in cement should be turned over all drains passing through walls, to prevent a possible settlement of the wall damaging the drain.

At those points where it becomes necessary to increase the size of the drain, the enlargement should be made by means of proper taper pipes or channels, the invert of the drain being laid at the same gradient throughout. When the enlargement does not occur in an inspection chamber, it should be made as shown in Fig. 17.

CHAPTER VIII.

INSPECTION AND INTERCEPTING CHAMBERS.

INSPECTION CHAMBERS OR MANHOLES:—Method of arranging the junctions of drains—The object of inspection chambers—Their size and construction—Arrangement of branch channels—Formation of benchings—Manhole covers—Inspection chambers formed with rock concrete tubes.

INTERCEPTING CHAMBERS:—Disconnection of drains from public sewer—Construction of intercepting chambers—A self-cleansing trap necessary—Remarks respecting the cleaning arm of the intercepting trap.

INSPECTION CHAMBERS OR MANHOLES.

IT is desirable that the junctions of all drains may be effected within a chamber specially constructed for that purpose, the portion of the drains and branches within the chamber being formed with open channels.

A series of such chambers should be so arranged that the whole of the drains within the system are accessible by means of drain rods. Any obstruction can then be readily removed from the drains, if necessary, without incurring the expense and inconvenience of opening up the drain. These inspection chambers also afford every facility for a periodical examination or testing of the drains when required.

A convenient size for ordinary inspection chambers is 3 feet by 2 feet 6 inches. This allows sufficient room for a man to work conveniently therein with drain rods. In places where the change of direction of the flow of sewage is at right angles to the previous direction, it is better to make the inspection chambers 3 feet square. When over 8 feet in depth, the upper portion of the chamber is contracted in size by constructing an arch at a height of about 5 feet above the concrete bottom, an access shaft, 2 feet 6 inches by 2 feet in the clear, being carried up from the soffit of the arch to the surface of the ground.

Figs. 18 to 21 show the details of construction for an ordinary inspection chamber through which a 6-inch main drain is passing, and receiving the discharge of three branch drains. The foundations for the chamber are of Portland cement concrete, the sides being built with white glazed bricks set in cement. If white glazed bricks cannot be used, on the ground of expense, the sides of the chamber may be built with carefully selected ordinary bricks of a hard impervious description. The sides are sometimes rendered in Portland cement, and afterwards limewhited.

The open channels should be formed with the best white glazed stoneware channel pipes, bedded and jointed to proper falls in the concrete bottom. All branch channels must be arranged to discharge *over* the edge of the main channel, in order to avoid any back-flow or wash. Branch channels should never be

56   SANITARY HOUSE DRAINAGE.

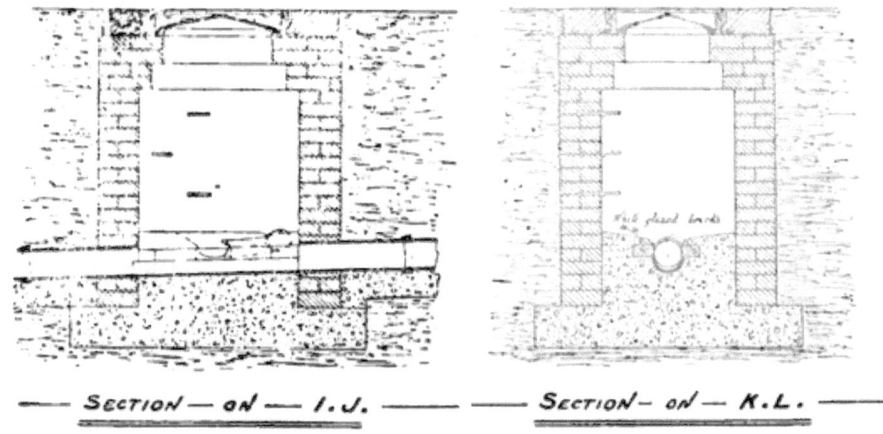

SECTION — ON — I.J.   SECTION — ON — K.L.

Fig. 18.   Fig. 19.

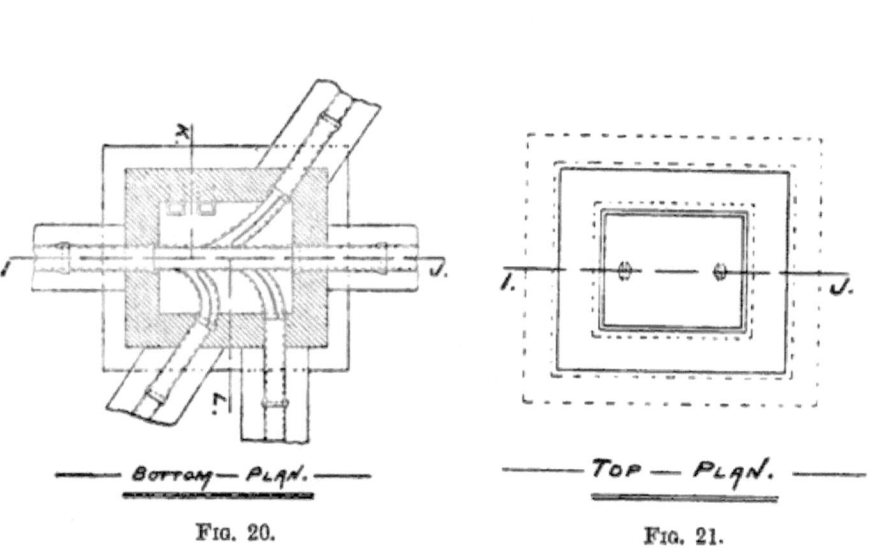

BOTTOM — PLAN.   TOP — PLAN.

Fig. 20.   Fig. 21.

INSPECTION CHAMBERS OR MANHOLES.

permitted to discharge directly opposite each other, as this is a great source of splashing, should the branches discharge their contents together. The proper arrangement is shown in Fig. 20. It is necessary to give the branch channels as large a sweep as possible, so that the direction of the flow of the branch may be gradually given the same direction at the point of discharge as that of the main channel. Open channel pipes of a flat sweep may be semicircular in section; but for channels having a fairly quick sweep, it is desirable that the bends be of three-quarter pipe section, as shown in

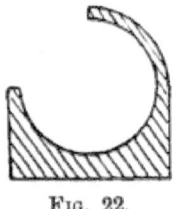

Fig. 22.

Fig. 22. These must be so laid that the *outer* edge of the bend is higher than the inner, in order that the discharge when passing round the bend may be folded over like a wave, and turned in the direction of the flow of the main channel.

The concrete bottom of the chamber must be *benched* or sloped to the same height as the top of the drain pipes discharging into the chamber, the benchings being formed so as to slope towards their respective channels, and neatly finished to a smooth face with cement. The slope of the benchings should be flat enough to permit a man to stand conveniently on them. It is a great improvement to lay a course of white glazed bricks along the sides of the main channel in the spaces between the branch channels, as shown in Figs. 3, 7, 19, and 28.

For convenience of access, the top of the chamber is covered with an iron cover and frame, and galvanised ladder irons (spaced about 12 inches apart) are built into the sides. For ordinary purposes, a cover 2 feet by 1 foot 6 inches will be found a convenient size, the sides of the chamber being gathered over, so as to reduce the opening to these dimensions. A suitable stone or concrete curb should be provided round each manhole cover. Where the inspection chambers are situated near dwellings, frequented paths, or, in confined areas, the manhole cover should be perfectly air-tight, and in such cases a "double" manhole cover may be used with advantage. Where drains are carried round the angles of buildings, it is desirable to construct a square manhole about 3 feet square, as shown in Figs. 23 to 26. Inspection chambers, circular on plan, are frequently constructed of "rock concrete tubes," as shown in Figs. 27 to 30. Where these are used for house drains, the lower portion of the chamber is 3 feet diameter, the upper portion being reduced to 2 feet 3 inches diameter by means of a specially made taper piece. Ladder irons are built into the sides of the chamber, convenient apertures being left in the tubes for that purpose.

INSPECTION CHAMBERS OR MANHOLES.

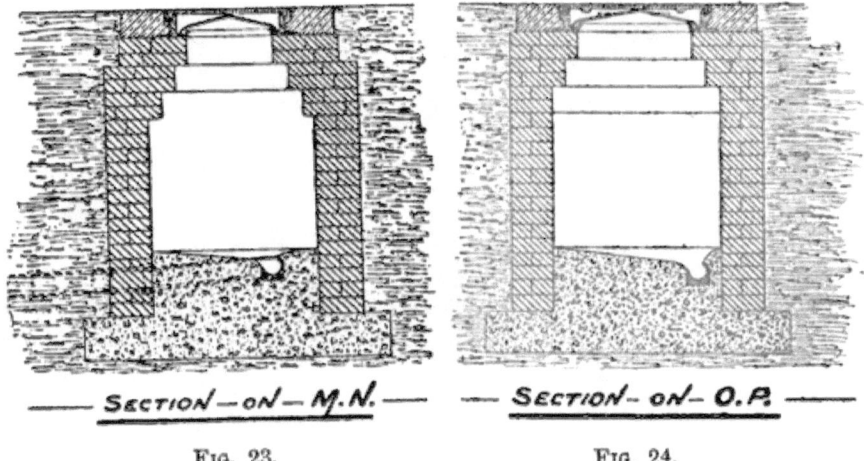

SECTION ON M.N.
Fig. 23.

SECTION ON O.P.
Fig. 24.

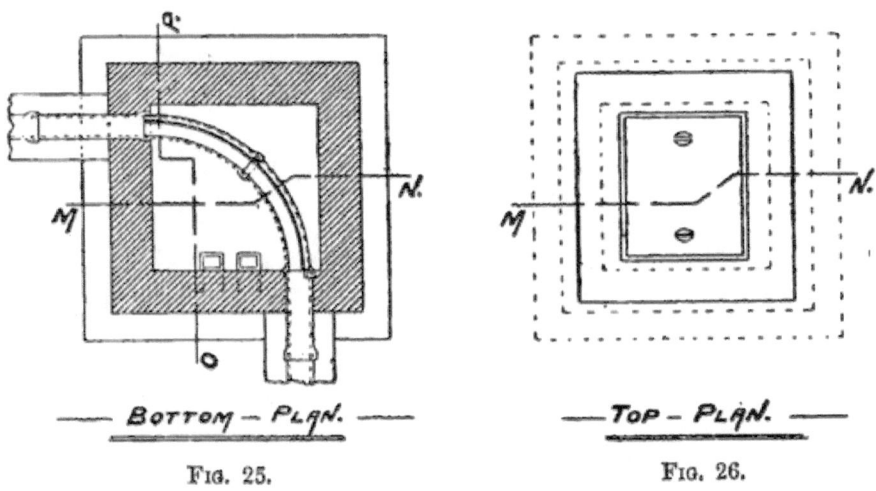

BOTTOM PLAN.
Fig. 25.

TOP PLAN.
Fig. 26.

60 SANITARY HOUSE DRAINAGE.

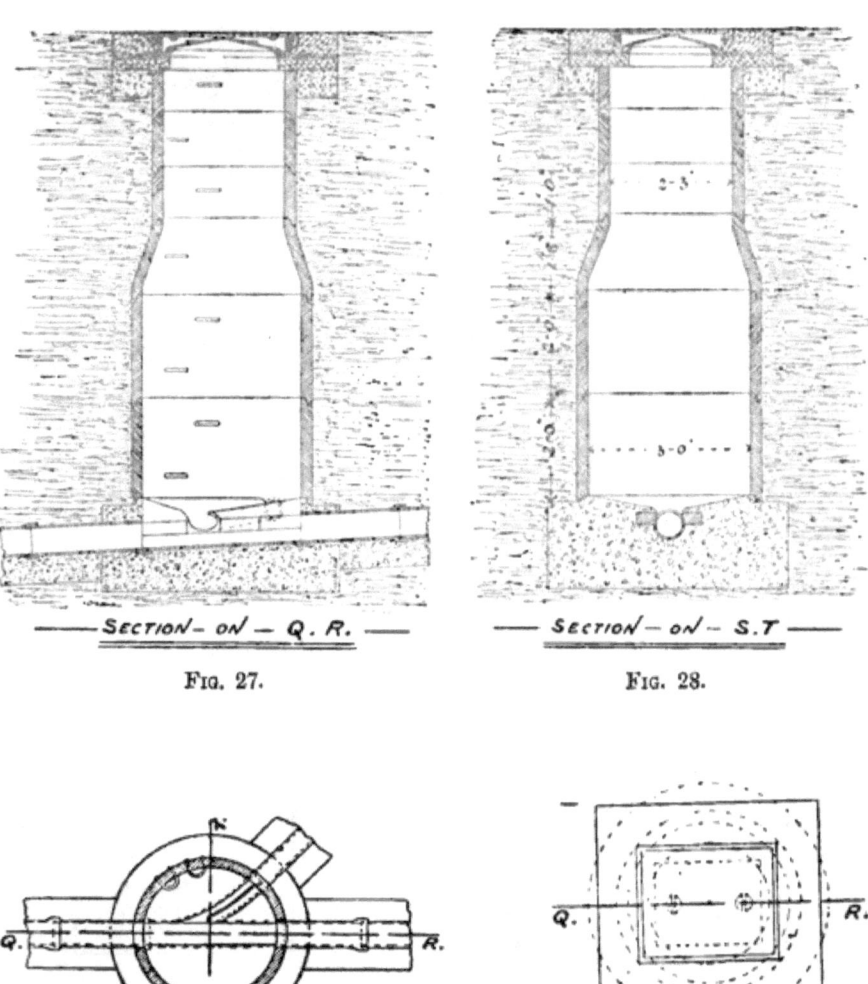

Section on Q. R.

Fig. 27.

Section on S.T.

Fig. 28.

Bottom Plan.

Fig. 29.

Top Plan.

Fig. 30.

Intercepting Chambers.

It has already been stated that the domestic drainage system must be properly disconnected from the public sewer at some convenient point near its junction with the latter. This disconnection is effected within a chamber constructed for the purpose, so as to admit of examination at any time. The general arrangments of an ordinary intercepting chamber have been shown in Figs. 2 to 5, and also in Figs. 6 to 9. The details of construction are similar to those described for inspection chambers, with the addition of an intercepting trap at the outlet of the drain, and adequate provision for the continuous supply of fresh air to the drainage system.

It is necessary to select a well-designed self-cleansing intercepting trap; otherwise the drain will be liable to constant stoppage at this point, and prove a continual source of annoyance and expense. The intercepting trap must be set perfectly level. To ensure this, it is usual to bed a piece of 3-inch York paving on the concrete, the upper surface of the slab being levelled to receive the trap (see Fig. 6). A cleaning arm fitted with a movable cap is provided to the trap, so as to allow of access to the drain on the other side of the water seal, if desired at any time. It is important, however, to make sure that this arm is properly sealed and made air-tight when the drain is not undergoing examination.

CHAPTER IX.

SOIL, VENTILATING AND WASTE PIPES.

SOIL PIPES :—Materials generally used—Lead soil pipes—Table of weights—Objections to lead soil pipes—Joints of lead pipes—Joint between lead and cast-iron pipes—Joint between lead and stoneware pipes—Advantages of cast-iron soil pipes—Table of weights—Preservatives—Joints—Cast-iron pipes to be fixed with holderbats or blocking pieces.

VENTILATING PIPES :—Materials used—Precautions to be observed in the use of cast-iron pipes—Galvanised wrought-iron pipes.

WASTE PIPES :—Air disconnection essential—Wastes to be trapped—Materials and sizes used for vertical wastes—Table of weights and thickness of drawn lead pipes—Sizes of lead waste pipes.

SOIL PIPES.

SOIL pipes are usually constructed of lead or cast iron. Stoneware pipes have sometimes been used for this purpose, but they are most unsuitable, owing to the numerous joints required and the difficulty experienced in making them thoroughly air-tight, in addition to their general unsightliness.

Wrought-iron and steel pipes put together with screw connections have recently been used for soil pipes in the United States. They are protected from oxidation by one of the preservative processes usually adopted for iron pipes, such as the Angus Smith or

SOIL PIPES.

Bower-Barff method. The advantages obtained by the use of these pipes are that they can be procured in long lengths, and are easily supported at the sides of the buildings owing to their comparatively light weight. This latter consideration is an important one where the soil and ventilating pipes are of great height, as in some of the lofty American buildings.

Where lead soil pipes are used they should be in 10-feet lengths, and of the description known as "hydraulic drawn," the lead being of sufficient thickness to weigh not less than 7 lbs. per superficial foot. It may be mentioned here that, to comply with the London County Council's regulations, all lead soil pipes must have a minimum substance of 7 lb. lead.

The following is a table of the weights of hydraulic-drawn lead pipe usually employed for soil and waste pipes, viz. :—

Weight of Lead, Soil and Waste Pipes per 10-Feet Length.			
Internal diameter of pipe.	6 lb. thickness or ·101 in.	7 lb. thickness or ·118 in.	8 lb. thickness or ·135 in.
2½ in.	41 lb.	48 lb.	55 lb.
3 ,,	49 ,,	57 ,,	66 ,,
3½ ,,	57 ,,	67 ,,	76 ,,
4 ,,	65 ,,	76 ,,	87 ,,
5 ,,	..	94 ,,	107 ,,
6 ,,	..	112 ,,	128 ,,

Soil pipes are usually 4 inches in diameter, and for ordinary purposes no larger size will be required. Even in a large institution or public building, a soil pipe of greater diameter than 4 inches will scarcely ever be found to be absolutely necessary.

The greatest drawback to the use of lead for soil pipes is the liability of the pipe to "creeping," or a gradual downward movement of the metal, due to its expansion and contraction with every variation of temperature. The only remedy for this is to secure the pipe to the wall at short intervals with heavy cast-lead tacks. The tacks should be spaced not more than 3 feet apart, preferably in pairs, strongly soldered to the soil pipe, and secured to the wall with stout pipe nails driven into the horizontal joints of the brickwork or masonry.

The joints between lead pipes should, in all cases, be properly made with a well-wiped solder joint. With regard to the connection between the lead soil pipe and the drain, the joint must be very carefully made so that it may prove permanently satisfactory.

Fig. 31 shows the usual method of forming the connection between a lead soil pipe and a cast-iron drain pipe. To receive the soil pipe the drain is turned up with a convenient socket bend, and firmly bedded in a block of concrete. A strong brass ferrule is passed over the end of the lead pipe and securely soldered thereto at its upper edge, whilst at the lower end the lead is well dressed over it and inserted in the iron socket. A few gaskets of yarn are then forced in, and the joint run with lead and caulked.

The connection between a lead soil pipe and a stoneware drain pipe is made in a similar way, neat Portland cement being substituted for the molten

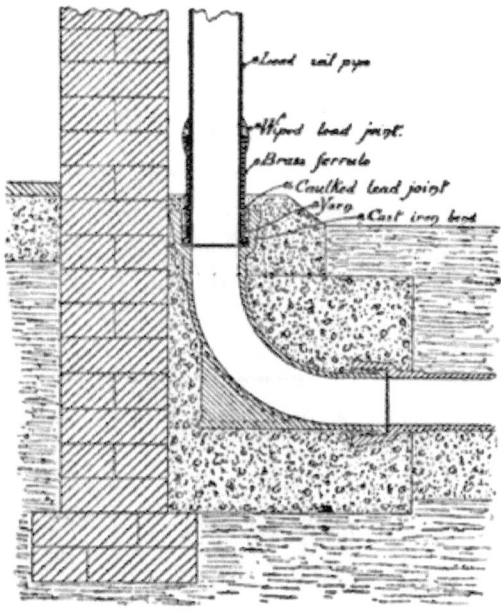

FIG. 31.

lead. Additional security may be given to the joint by afterwards surrounding it with 6 inches of concrete. When the bend to the drain is of iron it is better to have a "foot" cast on, as shown in Fig. 31, so that the bend may take a firm bearing on the concrete.

Cast-iron soil pipes are being largely used in the

place of lead soil pipes with satisfactory results, more especially as they can now be obtained with enamelled and perfectly smooth interior surfaces. They are also free from the "creeping" tendency to which lead soil pipes are subject.

Cast-iron soil pipes should be of good tough grey cast iron, and fulfil all the conditions previously described for cast-iron drain pipes, with the exception of the hydraulic test, as it is not necessary that they should be so strong and heavy as drain pipes. They are generally made in 6-foot lengths, and may be obtained with or without ears for fixing to the wall. The following table shows the weight of cast-iron soil and waste pipes, as required by the regulations of the London County Council:—

WEIGHT OF CAST-IRON SOIL AND WASTE PIPES PER 6-FOOT LENGTH.

Internal diameter of pipe.	Weight per pipe.
3 in.	40 lb.
$3\frac{1}{2}$,,	48 ,,
4 ,,	54 ,,
$4\frac{1}{2}$,,	62 ,,
5 ,,	69 ,,
6 ,,	84 ,,

Cast iron should be coated with some preservative to prevent oxidation. They may be "glass enamelled" inside, and painted externally after fixing, or coated with the "Dr. Angus Smith" process. Where the latter process is adopted, and it is required to paint them externally when fixed, the pipes must be given a coat of patent knotting before painting, so as to prevent the tar forcing its way through the painted surface.

The joints of all cast-iron soil pipes should be run with molten lead and well caulked. The pipes should be secured to the wall with iron holderbats or blocking pieces, built into the brickwork or masonry, so that the pipes may be blocked off about 2 inches from the wall. By this means sufficient space is given to allow of the pipes being painted all round, and the joints inspected at any time. Where pipes with ears are used, they should be blocked off at least 1 inch from the wall, by passing short pieces of iron pipe over each pipe nail before fixing.

Ventilating Pipes.

These may be of lead or cast iron, and the same precautions as regards the substance of the pipes, method of forming the joints, &c., should be adopted as already described for soil pipes. Cast-iron ventilating pipes should invariably be coated with some preparation to prevent oxidation, otherwise the interior of the pipe is liable to be choked with accumulations of iron rust, and thus become utterly useless for ventilation purposes.

Galvanised wrought-iron ventilating pipes in 6-foot

lengths, with sockets for running with lead and caulking, are now manufactured. They can be obtained with or without ears, and are useful for situations where the ventilating pipes are required to be of great height.

Waste Pipes.

Where lavatories and baths are situated on the ground floor, the waste pipes leading from them should discharge over a trapped gully. Complete air-disconnection is thus established between the fitment and the drain. At the same time the waste pipes themselves must be efficiently trapped close to the outlet of the fitment, in order to prevent the passage of air from the outside to the inside of the building, as all air entering in such a way would of necessity become vitiated by contact with the fouled surfaces of the pipe. When these fitments are on an upper floor, the waste pipes should discharge over the hopper-head of a vertical waste, which, in its turn, should discharge over a trapped gully. Being open to the air at both ends, the vertical waste will consequently be thoroughly ventilated.

In cases where a series of baths or lavatories are placed over each other, and discharge into one common vertical waste, it is usual for the waste pipe to be carried up above the eaves, the lower end discharging over a trapped gully.

The vertical wastes are constructed of lead or cast-iron pipes, with air- and water-tight joints, as described for soil pipes. The size generally varies from 2 inches to 3 inches in diameter, according to the quantity of waste water to be removed. Even for a large institution it should scarcely ever be actually necessary to employ a vertical waste larger than 3 inches diameter.

The usual weights of cast-iron or lead vertical waste pipes have already been given in connection with the table of weights of soil pipes. The following table gives the weight and thickness of drawn lead pipe suitable for the branch wastes from baths, lavatories, and other sanitary appliances. Those given under the heading of "strong" lead pipes are the weights required to comply with the regulations of the metropolitan water companies, and are therefore suitable for use as service pipes to the various fitments:—

Internal diameter or bore of pipe.	Middling.		Strong.	
	Thickness.	Weight.	Thickness.	Weight.
½ in.	·14 in.	4 lb.	·19 in.	6 lb.
¾ ,,	·15 ,,	6 ,,	·20 ,,	9 ,,
1 ,,	·16 ,,	9 ,,	·21 ,,	12 ,,
1¼ ,,	·18 ,,	12 ,,	·23 ,,	16 ,,
1½ ,,	·19 ,,	16 ,,	·22 ,,	18 ,,
2 ,,	·20 ,,	21 ,,	·23 ,,	24 ,,

WEIGHT AND THICKNESS OF DRAWN-LEAD PIPES PER YARD RUN.

No waste pipe should be less than 1¼ inch in diameter, and the outlet from the fitment to the waste—whether consisting of a perforated grating, valve, or plug—should in all cases be of sufficient effective area to allow of the waste pipe running "full bore," so that both trap and waste may be thoroughly cleansed at each discharge.

Waste pipes of the following sizes are recommended so as to provide a "quick" waste, and also to assist in some measure to flush the drains, viz.:—

Lavatory basin, 1¼ inch or 1½ inch waste.
Urinal basin, 1½ inch ditto.
Butlers' sinks, 1½ inch ditto.
Baths, 1½ inch or 2 inch ditto.
Scullery sinks, 2 inch ditto.

Where a lead or other safe is fixed under any sanitary fitment, it should be provided with a 1¼-inch diameter waste pipe discharging into the open air, and finished with a copper or brass hinged flap valve on the outside.

CHAPTER X.

MANHOLE COVERS.

MANHOLE COVERS:—Form and size of covers—Direct ventilating cover—Solid manhole cover—Substances used for sealing the joint between cover and frame—Solid cover with sunk top—Advantages of sunk top—Double manhole cover—Double manhole cover with sunk top.

EVERY inspection and intercepting chamber should be provided with a movable iron cover, so as to allow of ready access to the drains when necessary. The cover should be as small as possible, provided that it is sufficiently large to permit a man of ordinary size to pass through conveniently. It may be either circular or rectangular on plan, but for ordinary domestic purposes rectangular covers are most frequently used, as the curb or paving can be more easily fitted to them.

Manhole covers having a clear opening of 2 feet by 1 foot 6 inches, or 1 foot 8 inches diameter, are a very convenient size for general use.

Fig. 32 shows a "ventilating" manhole cover, suitable for situations where direct ventilation to the drains is required. A dirt box is provided under the perforated cover to intercept the passage of dirt, stones, sticks, or other foreign substances which might otherwise enter the drain through the openings. The dirt box should be periodically taken out and emptied.

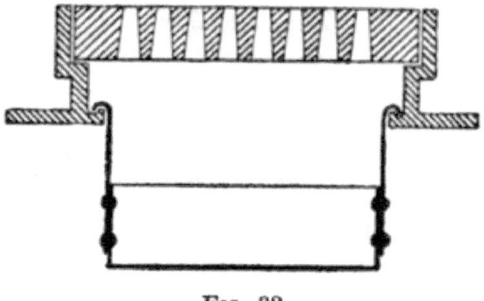

Fig. 32.

The common form of solid manhole cover is shown at Fig. 33. The frame is formed with a groove into which the cover is fitted. The groove may be filled with oil, glycerine, or indiarubber packing, to prevent the passage of vitiated air from the drain.

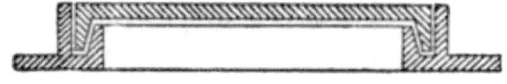

Fig. 33.

For positions where the large surface of iron plate would be unsightly, as in yards, areas, &c., a solid sunk cover similar to that shown in Fig. 34 may be used, the centre portion being filled with concrete, tiles, wood paving, or other material, so as to accord with its immediate surroundings.

In confined situations where it is necessary that increased precautions must be taken against the possible escape of any impure air from the drain, a

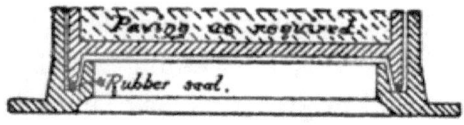

Fig. 34.

"double" manhole cover should be used, an example of which is shown at Fig. 35. The detached inner cover is arched so as to allow the moisture from the drain to condense on its under side and run down into

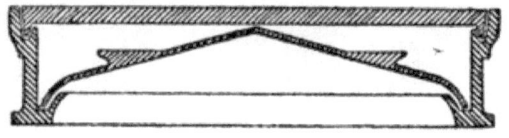

Fig. 35.

the groove, thus providing a water seal at that point. The space between the inner and outer cover may be filled with sand, charcoal, or other suitable deodoriser, if desired.

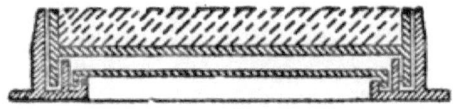

Fig. 36.

Fig. 36 shows another form of "double" cover designed to admit of the centre portion being filled in with paving material similar to that of the adjacent surface. The grooved joint of the inner cover is sealed with oil or glycerine, and the space between the two covers filled with sand, charcoal, or other material. The joint of the outer cover is deeply sealed with water from the rains, or washing of the floors, but in dry weather oil or glycerine may be used.

CHAPTER XI.

TRAPS.

INTERCEPTING TRAPS:—Best form of trap or seal—Remarks respecting traps with water seal—The essentials of a good intercepting trap—Desirability of a cascade or waterfall action—Traps rendered more self-cleansing when formed with a contracted throat—Circumstances under which traps without cascade action may be used.

PLUMBERS' TRAPS:—Internal fitments to be trapped—D-traps objectionable—Defects of bell traps—Siphon traps—Anti-D-traps—Water seal of traps broken by evaporation, momentum, and siphonage—Arrangement and size of anti-siphonage pipes.

INTERCEPTING TRAPS.

IN all cases where a trap or seal is required for drainage or sanitary purposes, it should be of the simplest possible form. It must be automatic in action, and free from any liability to get out of order. A trap or seal dependent only on some mechanical arrangement should be avoided. The best and simplest form of trap at present devised is that known as the siphon trap with a water seal. The least effective water seal that should be permitted in any trap is $1\frac{1}{2}$ inches.

It may be incidentally mentioned that even a deep water seal cannot theoretically be considered as constituting an absolute or perfect trap under all circumstances; but it fulfils every practical sanitary requirement, provided that the water seal is kept intact. It has been frequently demonstrated that minute quantities of certain gases, as ammonia, sulphuretted-hydrogen, chlorine, &c., may, under very favourable conditions, pass through the water seal to the other side of the trap, but any such gas that has passed is infinitesimal in volume and quite innocuous. So far as it has yet been ascertained, no disease germs are able to pass through the motionless water seal of a good siphon trap.

A well-designed and effective intercepting trap for drainage purposes, in addition to being provided with a good water seal to prevent the passage of sewer air or gas from one side of the trap to the other, must be thoroughly self-cleansing. It should be free from any sharp angles, and perfectly smooth and even throughout, so that the surfaces may not become fouled, nor the trap liable to stoppages at any part of it when ordinary sewage matters are passing. The area of the exposed water surface on each side of the water seal should be as small as practicable, to prevent any rapid evaporation of the water constituting the trap or seal.

It is also desirable that the trap should be so constructed as to provide a difference of level of at least 3 inches between the inlet and outlet. By this means a "cascade" or waterfall action takes place at the inlet, and increased velocity is given to the flowing sewage, so that it may more effectually overcome the frictional resistance necessarily offered by the form of the trap itself. In order to still further obtain an increased cleansing and scouring action through the trap, it is frequently formed with a slight contraction at the throat or lower portion of the passage of the trap. Fig. 37 is a section through a 6-inch intercepting trap,

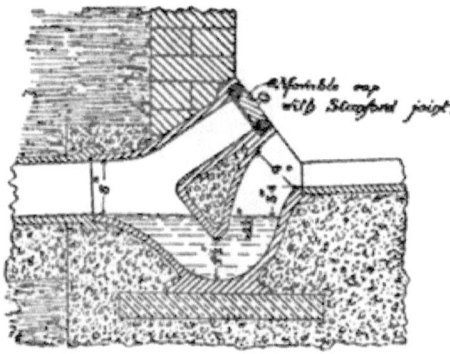

FIG. 37.

of a good self-cleansing type, having a 3-inch cascade, $2\frac{1}{2}$-inch water seal, and a contraction of $1\frac{1}{2}$ inches at the throat.

In situations where only a slight general fall can be obtained for the drainage system, it is more important that all the available fall may be reserved for the drains themselves rather than utilise a portion of it for a waterfall action to the intercepting trap. Under such circumstances it is necessary to provide a trap in which the inlet and outlet are on the same level, as shown in Fig. 38, the cascade or waterfall being omitted. A cascade action intercepting trap should, however, be used in all places where sufficient fall is obtainable.

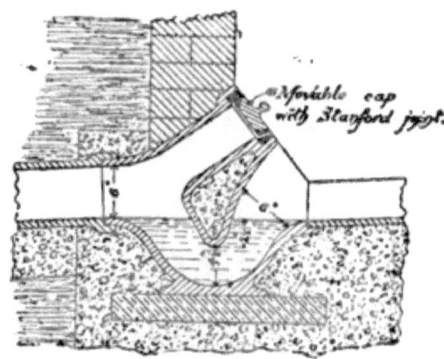

Fig. 38.

Plumbers' Traps.

All internal sanitary appliances, such as water-closets, sinks, baths, lavatories, &c., must be provided with some form of trap or seal near the outlet of the fitment. This is necessary even if the discharge is carried directly into the open air through a short waste pipe, in order that the external air may not be permitted to enter the building through the pipe. The interior of any waste pipe is more or less fouled by constant use, and any air passing through it becomes contaminated and rendered unfit for respiration.

The essentials of a good self-cleansing trap having been given in the foregoing pages, need not be repeated.

Plumbers' traps were at one time almost invariably of the form known as the "D trap," the general arrangement being as shown in Fig. 39. It is scarcely necessary to say that such a trap is insanitary in principle and should never be used. The interior becomes

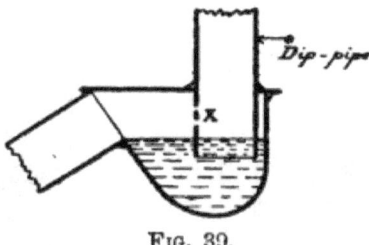

Fig. 39.

coated with filth which no amount of flushing will remove, and should the sides of the dip pipe become corroded and perforated, as indicated at X, Fig. 39, the seal of the trap is entirely destroyed, the vitiated air

Fig. 40.

having free access into the building, though no outwardly visible indication of the danger is given.

Another most insanitary form of trap, at one time greatly used for sink wastes, was that known as the "bell trap" (see Fig. 40). It possesses a very shallow water seal, becomes thickly furred with passing waste matters, and is constantly liable to stoppages.

A simple and satisfactory form of trap is the ordinary "siphon trap," shown in Figs. 41 and 42.

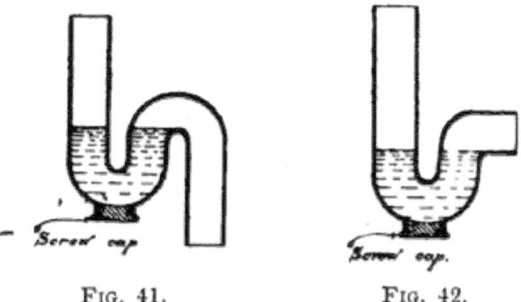

FIG. 41.   FIG. 42.

Another excellent type is that known as the "anti-D trap," made of refined cast lead (see Fig. 43). Both forms are self-cleansing, particularly the last named,

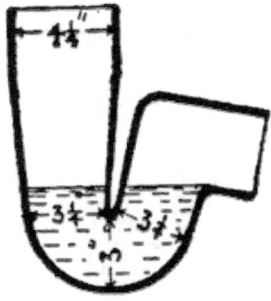

FIG. 43.

as it is slightly contracted at the throat, in order to obtain an increased scouring action through the lower portion of the trap.

For pipes of small diameter, as the wastes from

lavatories, sinks, &c., the traps should be provided with a brass screw cap for inspection purposes, as shown in Figs. 41 and 42.

Lead siphon traps may be either hand-made, cast, or hydraulic drawn. Where lead traps are used, they should either be of the " anti-D " type, or those known as the Du Bois hydraulic drawn, perfectly smooth

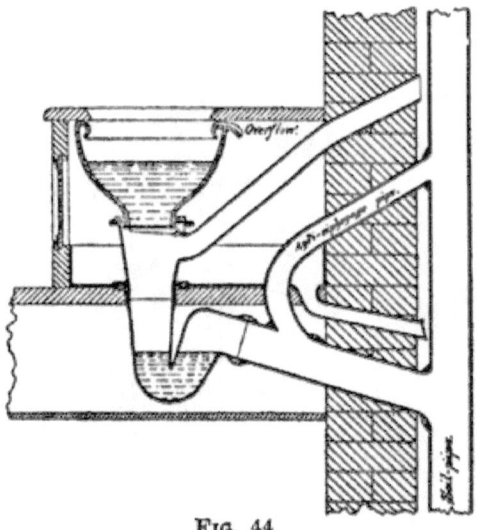

FIG. 44.

inside, and free from flaws of every description. The latter can be obtained in lead varying from 6 lb. to 8 lb. substance.

All water traps are liable to become unsealed by evaporation. The seal may also be broken by the water in the trap being " siphoned " or " momentumed " out. The water is said to be momentumed out when the body of water, which would otherwise be retained

82   SANITARY HOUSE DRAINAGE.

to form the seal of a trap, is carried out by the momentum or impetus due to its own mass and velocity. To

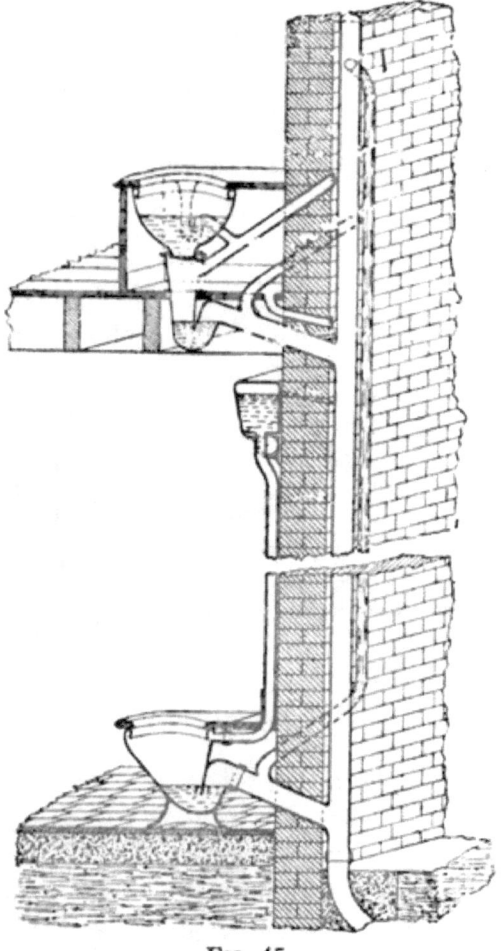

FIG. 45.

prevent siphonage of the water seal, all internal traps should be provided with an air pipe near the top of the trap. This is known as an "anti-siphonage pipe." For scullery sinks, lavatories and similar fittings, the anti-siphonage pipe must be taken direct into the open air, but for water-closets and housemaids' sinks such a method is not convenient, for it would necessitate the anti-siphonage pipe being carried above the eaves in the same manner as the soil pipe. The same object, however, is attained by connecting the anti-siphonage pipe with the soil pipe, as shown in Fig. 44. Where two or more closets are arranged vertically and discharge into the same soil pipe, the anti-siphonage pipes must be connected by means of a vertical pipe, and carried *above* the branch from the topmost closet before being joined to the soil pipe. (See Fig. 45.) The vertical anti-siphonage pipe should be fixed *outside* the building as shown.

For water-closets and housemaids' sinks, a 1½-inch or 2-inch diameter anti-siphonage pipe will generally be found sufficiently large, and 1-inch or 1¼-inch diameter for lavatories, scullery sinks, &c. The anti-siphonage pipe must not be fixed on the crown of the trap, but a little distance beyond, so as to avoid the liability of becoming choked (see Fig. 45).

Where a perforated grating is fixed to the inlet of the siphon trap, as for scullery sinks, &c., the entrance to the trap should be enlarged, so that the effective discharging capacity of the grating may at least be equal to that of the inlet of the trap. The enlargement should take the form of an inverted cone.

CHAPTER XII.

SURFACE GULLIES. KITCHEN AND SCULLERY SINKS.

SURFACE GULLIES:— Description of gully to vary according to its situation—Storm-water gullies to be trapless—Rain-water shoes as an alternative—Necessary precautions in the use of trapless gullies—Gullies for foul drains to be trapped—Arrangement of gully with cleaning arm to branch drain—Difficulty of removing greasy liquids—Construction of grease-traps—Objections to grease traps—Advantages of a flushing-rim grease gully with flushing cistern.

KITCHEN AND SCULLERY SINKS:—General arrangement and construction—Anti-siphonage pipe to be provided—Sinks receiving greasy liquids should discharge into flushing-rim gullies.

SURFACE GULLIES.

BEFORE selecting a surface gully for any particular situation, it is necessary to know the nature of the waste liquids that will usually be discharged into it, so that an efficient gully may be provided for each particular situation. A gully that would be suitable to receive the waste water from a roof or yard enclosure might be found ill adapted to receive the various greasy liquids from a scullery sink.

All gullies receiving storm water only should be *trapless*. For this purpose a trapless gully, as shown in Fig. 46, may be used. Any dirt, sand, or other débris from the roof is retained by the gully, and may be periodically removed, instead of being allowed to enter the drain. Similar trapless gullies may also be obtained fitted with a small silt bucket, for the easier removal of débris. A rain-water shoe, as Fig. 47, may be used in the place of a *trapless* gully, if preferred. It must be distinctly understood that *trapless*

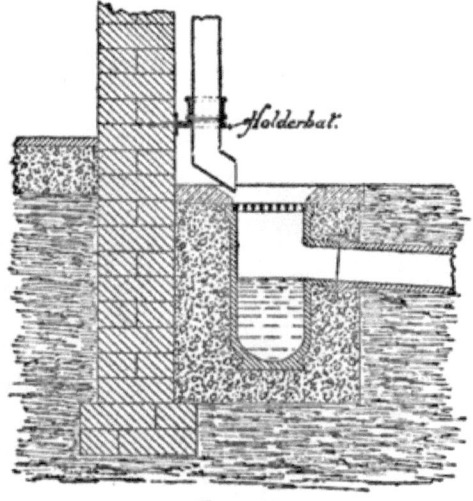

Fig. 46.

gullies or rain-water shoes can only be used in situations where the storm-water drains have been properly disconnected from the foul drainage section, as already described.

All surface gullies directly connected with the foul drainage system must be securely *trapped*. The gully trap should be thoroughly self-cleansing, and possess a good deep water-seal, so as not to become easily untrapped by evaporation.

When the branch drain from a trapped gully is not connected to the main drain in an inspection chamber, it is desirable to provide a ready means of access to the branch at some point near the gully. This is sometimes done by fixing a junction near the gully trap in such a manner that the arm is placed in a vertical

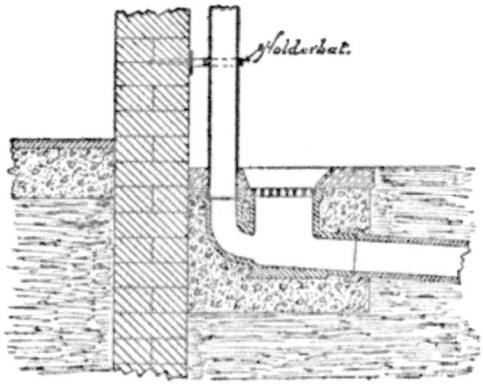

FIG. 47.

plane with the drain, so that it may be used as an inspection eye and cleansing arm if required. When not in use, the inspection eye is closed by means of an earthenware stopper, and afterwards covered with an iron or stone cover. A much more simple and satisfactory method to adopt in such circumstances is to provide a gully trap with cleaning eye attached, as shown in Fig. 48.

For gully traps receiving greasy liquids, such as

SURFACE GULLIES.

those required for scullery sinks, &c., many different expedients have been tried—with more or less success—to overcome the difficulty experienced in dealing with the removal of greasy liquids in a simple and efficient manner. The particles of liquid grease, becoming chilled as they reach the gully, are congealed and deposited within and around the sides of the trap. Unless special precautions are taken, these solid particles of grease eventually enter the branch drain, where, being unaccompanied by a sufficient quantity

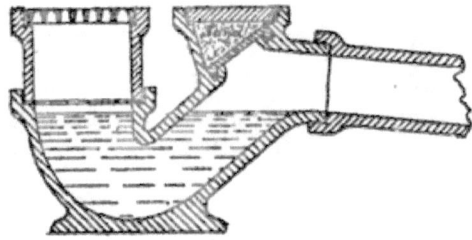

FIG. 48.

and force of water to carry them direct into the sewer, they gradually accumulate on the sides until the drain becomes thickly furred, resulting eventually in a complete stoppage.

It will generally be found that the ordinary discharge from a scullery sink waste is not sufficiently powerful to prevent the particles of grease being deposited on the sides of the drains and common form of trapped gully; but it is absolutely necessary to prevent such deposition, if the drains are to be kept in a permanently sanitary condition.

Until comparatively recently, it was usual to fix what is known as a "grease-trap" in all situations where liquids of a greasy nature might be expected. This consists essentially of a cast-iron or stoneware basin or receiver, so designed as to retain any fatty particles that may be discharged into it, whilst allowing the waste water—after being freed from grease—to enter the drain. The solidified grease must necessarily be removed by hand from the grease receiver at stated intervals.

Fig. 49 shows a grease-trap which is frequently used. The greasy water being discharged into the receiver is suddenly cooled by contact with the comparatively large volume of water retained there, and the now congealed particles of grease, being of light specific gravity, rise to the top of the water in the receiver and remain there until removed, the waste water flowing under the grease into the drain. Where the grease receiver is fitted with an air-tight cover, it is desirable that in all cases it should be thoroughly ventilated by means of a fresh-air inlet and a foul-air outlet. The foul-air extracting pipe should discharge into the air at some convenient point where it would not prove obnoxious or injurious to health.

Grease-traps may be considered as insanitary in principle, for the great aim of domestic sanitation should be the immediate removal of all impure matters from the house drains. The necessity should not, therefore, arise for the collection of grease or any other description of sewage, which must be periodically removed by hand from the drains or traps in a more or less advanced state of decomposition.

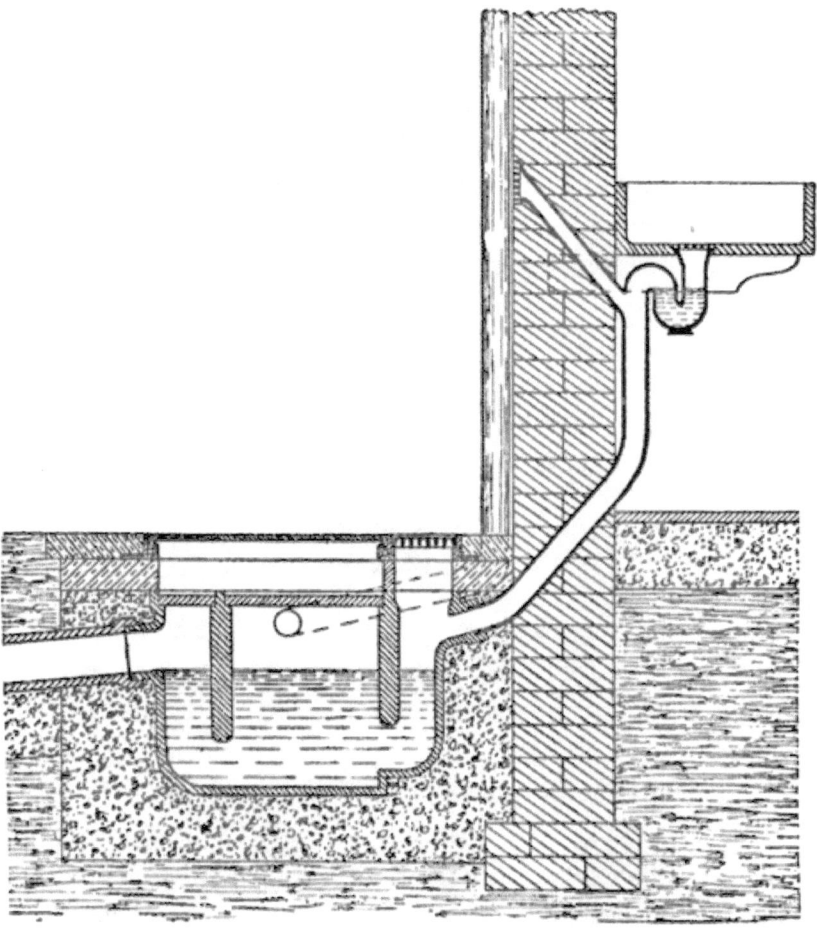

Fig. 49.

Greasy liquids may be most satisfactorily dealt with if allowed to discharge into a properly constructed flushing-rim grease-gully, provided with an automatic flushing cistern.

Fig. 50 shows a well-known form of flushing-rim grease-gully. This contains a sufficient volume of water to chill any grease that may be discharged in solution from the sink, the congealed fatty particles rising to the water surface in the gully. The flushing-rim of the grease gully is connected with an automatic flushing cistern, the water from which, being discharged with great force, breaks up the solidified grease within the

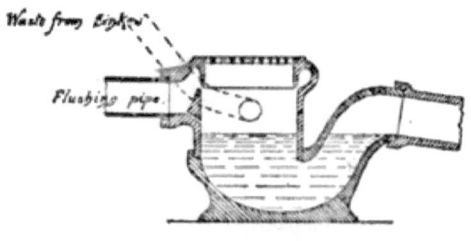

Fig. 50.

gully; the particles are carried through the trap and entirely removed from the drain, whilst at the same time the gully and drain are thoroughly flushed and cleansed.

For ordinary purposes it will be found that a 30 or 40-gallon automatic flushing tank discharging once a day is sufficient to cleanse the grease-gully and branch drain; but for hotels and other large establishments it will be necessary to flush the gully at least twice a day.

The waste from the sink may discharge *over* the grease-gully (as shown in dotted lines in Fig. 51) instead

SURFACE GULLIES. 91

of *below* the flushing-rim, if desired. The objection to such an arrangement is that the grating and portion of

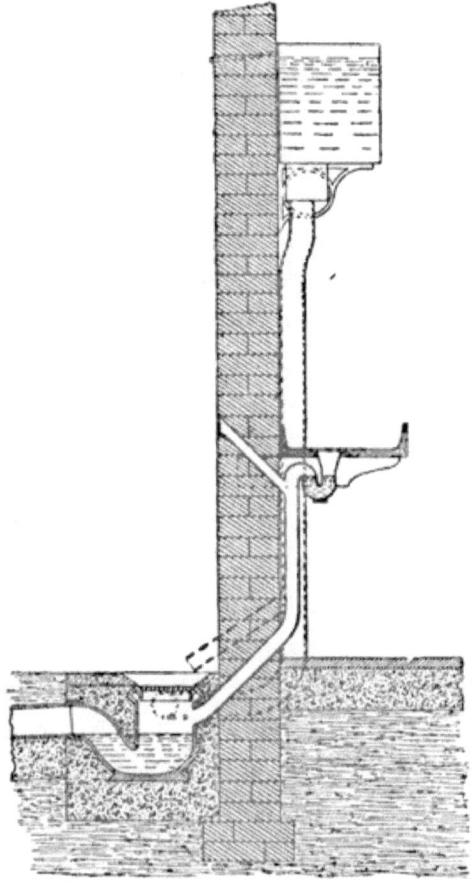

Fig. 51.

the gully above the flushing-rim are liable to be covered with a grease deposit, which of course cannot be removed by the periodical flush sent into the gully.

KITCHEN AND SCULLERY SINKS.

These should be made of strong white glazed stoneware, free from fire-cracks or flaws of any description. A strong brass grating about $3\frac{1}{2}$ inches diameter should be securely fixed to the outlet of the sink, with a 2-inch diameter trap, having a cone inlet immediately underneath, and connected to a 2-inch diameter lead waste pipe (see Fig. 51). The bottom of the trap should be provided with a screw cleaning eye for cleansing and inspection purposes. All sinks should be fixed on iron brackets instead of brick piers or wooden bearers, so as to avoid forming corners and resting places for dirt. To prevent any risk of siphonage, an air-pipe should be fixed near the top of the trap, and discharge into the open air. The outer end of the pipe should not be placed at a lower level than the top of the sink, and may be provided with cross wires soldered thereto, or a brass perforated hinged grating may be let in flush with the face of the wall, as shown.

Kitchen or scullery sinks should discharge into a flushing-rim grease-gully, as previously described, the gully and drain being kept free from any accumulations of grease by means of frequent flushing from an automatic flushing tank. Should it not be practicable to adopt flushing arrangements for this purpose—owing to an intermittent or insufficient supply of water—a grease-trap may be provided, as shown in Fig. 49. The removal of the grease should be carried out systematically and at stated intervals, whilst at the same time all parts of the grease-trap should be thoroughly flushed and cleansed.

CHAPTER XIII.

WATER-CLOSETS.

WATER-CLOSETS :—Water-closets to be trapped and the soil pipe ventilated—The essentials of a satisfactory closet fitment—Objections to pan closets—Construction of plunger closets—Valve closets—Conditions governing the construction of a good valve closet—Defects of long and short hopper closets—Objections to wash-out closets—Merits of a well-designed wash-down closet—Constructional details relating to wash-down closets—Siphonic-action closets.

CONSIDERABLE care must be exercised in the selection of a thoroughly sanitary type of water-closet, for the provision of a defective appliance may be attended with the most serious consequences. Of late years great improvements have been made in their general design and details of construction; but large numbers of water-closets are being manufactured at the present time which cannot be considered as complying with ordinary sanitary requirements.

All water-closets should be securely trapped close to the outlet of the fitment, so as to prevent the possible entry of impure air or gases from the soil pipe. This precaution must be supplementary to the provision of a continuous circulation of air at the junction of the closet with the soil pipe, as already mentioned in connection with the ventilation of the drains. The water-seal or trap may form part of the closet fitment itself, or may be fixed separately near its outlet. In either case the efficiency of the trap will be here considered as forming a component part of the sanitary efficiency of the closet as a whole.

For the production of a satisfactory and sanitary water-closet, it is necessary that the basin should be capable of retaining within it a volume of water affording sufficient area and depth for the complete immersion of fæcal deposits whilst the closet is being used, and the contents of the basin should be capable of being entirely conveyed through the trap into the drain by the discharge of water into the basin; at the same time the basin and trap should be recharged with clean water to permit of the proper use of the closet when next required.

Every portion of the closet, including the trap, must be thoroughly self-cleansing, with all its surfaces perfectly smooth, and having no tendency to become fouled by the passing fæces. A satisfactory closet should also be simple in construction, having no intricate working parts which may be liable to get out of order under ordinary use.

Although the type of closet known as the "pan closet" is considered to be practically extinct, yet notwithstanding all that has been written and said about it, the pan closet is still being made, and—presumably—used. For this reason a slight description of its characteristics are given. The general arrangement is shown at Fig. 52, and consists of a basin with a movable copper pan under, by which a small quantity of water is retained at the bottom of the basin. The pan is pivoted within a large iron container or receiver. On pulling a handle the contents of the pan are discharged into the receiver, a portion of which passes down the dip-pipe into the D-trap below, and from thence into the drain, the remainder being splashed

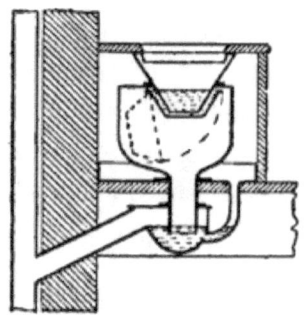

Fig. 52.

over and adhering to the sides of the receiver and D-trap. This gradually accumulates until both become thickly coated with decomposing filth. As a consequence, the apparatus becomes permanently insanitary, and each time the pan is emptied quantities of impure air and gases escape from the container into the room. A more insanitary form of closet could scarcely be devised.

A class of closet known as the "plunger" or "plug" closet was at one time extensively used (see

Fig. 53). A satisfactory area and depth of water is retained within the basin by means of the plug or plunger P. The basin is capable of being flushed in the usual manner, but the plunger chamber C is not self-cleansing; the sides of the chamber and plunger are liable to become furred with fæces, the deleterious gases from which escape into the apartment at the opening for the handle. This class of closet is also some-

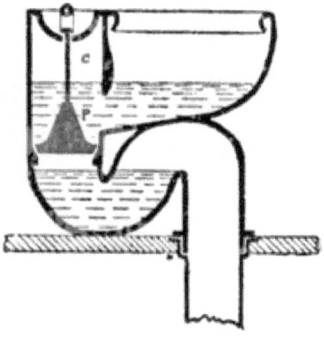

FIG. 53.

times fixed without a siphon trap between the soil pipe and the basin, and is then known as a "trapless" closet (see Fig. 54). In this case the only safeguard for preventing the entry of impure air from the drains into the house is that afforded by the plunger and the water above it. Should a small piece of paper or other substance prevent the plug from resting tightly upon its seat, the water above gradually escapes into the drain, and impure air is then free to enter the building —it may be with disastrous results.

The "valve" closet is a very satisfactory form of closet for domestic use, provided that a *really* good and well-made appliance is obtained. Many of those now being extensively sold will be found on examination to be faulty in design or workmanship, though they may have the questionable merit of being cheap in their first cost. A good valve closet is necessarily expensive, owing to its mechanism, and unless it is intended to use only a high-class valve closet, fixed in a complete and proper manner, they are better avoided altogether.

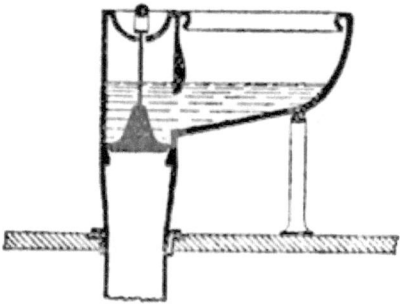

Fig. 54.

The general construction of a valve closet is shown in Fig. 44, and consists of a glazed earthenware basin, with an iron valve box under, and directly connected to a lead siphon trap secured to the soil pipe. One of the great advantages of this type of closet is that a large volume of water, having ample depth and surface area, is retained within the basin. In fact, the surface area of the water is generally greater than the opening in the seat, so that all fæcal deposits fall directly into the water pool without soiling the sides of the basin, or allowing unpleasant odours to be given off from the exposed fæces. After use, the basin is completely emptied of its contents, and the apparatus, including the trap, is well flushed, whilst a sufficient quantity of clean water is retained within the basin for its satisfactory use on the next occasion.

Amongst the details connected with the selection and fixing of an efficient valve closet, the following may be noticed, viz. :—

1. All valve closets must be provided with a *self-cleansing* siphon trap of good construction, immediately under the valve box.

2. A ventilating pipe (usually known as a puff pipe), must be provided to the valve box and carried through an outer wall into the open air.

3. An anti-siphonage pipe should be provided on the soil pipe side of the closet trap. For a single closet this may be connected to the soil pipe, as shown in Fig. 44, but in the case of a vertical series of closets the anti-siphonage branch from each closet is connected to a vertical pipe and afterwards to the soil pipe *above* the highest closet, as previously described and shown in Fig. 45.

4. A lead safe should be provided under each closet, having a waste pipe discharging into the open air, the outer end being provided with a brass or copper hinged flap.

5. The overflow from the closet basin should dis-

charge on to the lead safe beneath, as shown in Fig. 44. Where this plan is adopted, slops should *never* be emptied into the closet, otherwise there is great danger of organic matters being allowed to overflow and decompose within the lead safe should the closet be carelessly used.

Whenever there is any risk of the valve closet being used as a slop closet or housemaids' sink, the following arrangement should be substituted. The overflow pipe should be *trapped* and discharge into the ventilating pipe from the valve box, as shown in Fig. 45. It should never be allowed to discharge *directly* into the valve box. The overflow pipe must also be arranged to receive a small quantity of clean water on each occasion that the closet is flushed, so that the water seal of the overflow trap may be kept intact, and the whole of it should be easily accessible for cleaning purposes if required.

6. Where the local circumstances are such that a water-waste preventing cistern must of necessity be used in place of the ordinary supply valve and bellows regulator usually attached to the closet under the seat, the waste-preventing cistern should be capable of giving a "flush" and "after flush," so that the basin may be left with a proper quantity of water within it.

7. The enclosure to the closet should not be "fixed" or screwed down. The seat should be hinged, and the front or riser capable of being readily removed, so that the whole space may be properly cleaned from time to time.

Attention to these and similar details is necessary if an effective and satisfactory valve closet is desired.

The "hopper" form of closet, in its most primitive form, consists of an earthenware basin with a trap

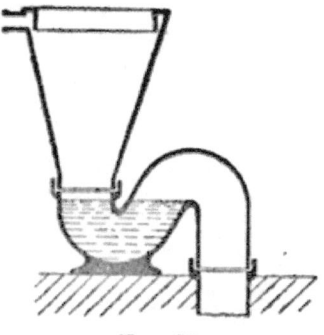

FIG. 55.

immediately under the outlet. The basin takes the shape of an inverted cone or "hopper"— hence the name. Hopper closets are known as "long" and

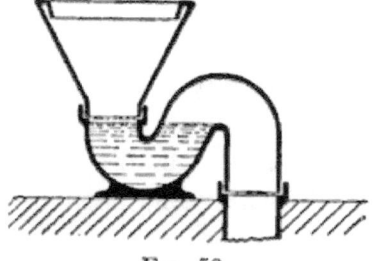

FIG. 56.

"short" hoppers, according to the depth of the basin. Figs. 55 and 56 are sketches of a "long" and "short" hopper closet respectively. The "short hopper" closet is frequently used for servants' closets and cottage property. It is cheap and simple in construction, but cannot be considered as complying with modern sanitary requirements. The surfaces of the basin are easily fouled and insufficiently self-cleansing, even when a flushing-rim is provided in lieu of the usual flushing-arm. The exposed water surface at the bottom of the basin is quite insufficient; whilst the area of the sides of the basin is so comparatively large that it is almost impossible to use the closet without soiling the sides.

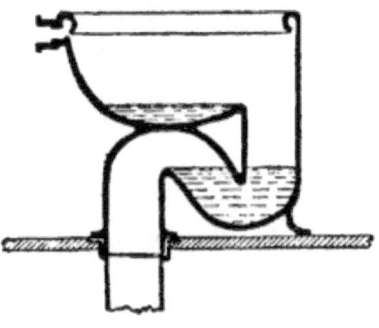

FIG. 57.

The result is that the ordinary two-gallon water flush given to these closets is quite incapable of overcoming the tenacious adherence of the fæces to the sides of the basin, and, under normal conditions, the closet permanantly remains in a more or less foul and insanitary state. With regard to the "long hopper," it will be noticed that the same defects are apparent, but in a much greater degree.

Some few years ago a class of closet known as the "Wash-out" was introduced, and at one time most extensively used (see Fig. 57). The closet is designed to allow of a shallow pool of water always remaining at the bottom of the basin for the reception of fæcal deposits. The surface area of the water pool is satisfactory, but the depth quite insufficient to permit the fæces being properly immersed without soiling the bottom and lower portions of the sides of the basin. The ordinary flush of water rarely removes all traces of the fæces which have thus soiled the basin. The force of the flush is also in a great measure expended in

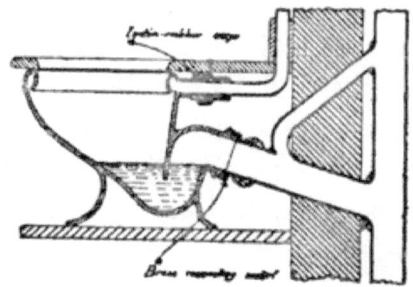

Fig. 58.

surmounting the weir formed in the basin, leaving insufficient energy to properly cleanse the outlet and trap under. As a consequence, instead of the closet trap being charged with clean water, it frequently retains a proportion of decomposing fæcal matters, whilst the sides of the basin outlet become gradually furred with the passing fæces, so that the whole appliance is rendered most insanitary.

Another form of closet, known as the "wash-down" closet, is shown at Fig. 58. This is essentially a modification of the short hopper and trap, and when of good design is a very satisfactory closet for general purposes. A fairly large water surface may be obtained within the basin, together with sufficient depth for the immersion of the fæces, whilst the continuously sloping sides of the basin allow the full force of the flush being brought to bear upon the trap with a thorough scouring and cleansing action.

In many of the wash-down closets now being manufactured, the water surface in the basin is far too small for sanitary efficiency, and they are consequently very little better than the common form of short hopper closet in that respect. A well-designed wash-down closet should provide a water pool of good depth and ample surface, together with a thoroughly self-cleansing trap. The back of the basin should be nearly vertical, whilst the outlet of the trap should be arranged not only to allow the connection with the soil pipe being properly and easily made, but to admit of it being at all times accessible for examination. It is desirable that wash-down closets may be of the pedestal type, in order to give adequate support to the upper portion of the fitment, and to allow the seat to rest directly upon the closet without the use of seat brackets. They should be fixed without any enclosure, the whole of the apparatus being exposed to view in order that it may be thoroughly cleansed in every part, no hidden corners being left for the gradual accumulation of dirt.

The description of closet known as the "siphonic-action" closet is of comparatively recent introduction. Fig. 59 shows the general arrangement of this type of closet. The outlet of the basin takes the form of a bent tube or siphon pipe, the long leg of which is connected to a trap (or, in some instances, a weir) attached to the soil pipe. On flushing the closet, siphonic action is set up, and the contents of the basin are siphoned or drawn out, instead of being thrust out by the force of the water flush. Nearly all manufacturers of siphonic-action closets adopt different arrange-

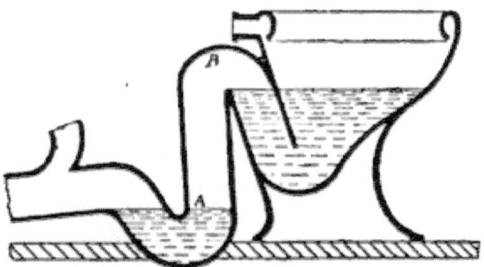

FIG. 59.

ments by which the air is expelled from the siphon pipe in order to set up siphonic action, this object being oftentimes attained in a highly ingenious manner. In one well-known closet of this description, a small puff pipe is fixed just above the water seal of the trap (at A, Fig. 59) through which the air is expelled by means of the rapid discharge of water (at B) down the long arm of the siphon pipe. In some forms of "siphonic-action" closet, waste water or slops must not be discharged into the closet for fear of setting up siphonic action, and so leaving the closet with little or no effective water seal between the soil pipe and the fitment.

By the use of a siphonic-action closet, ample depth of water is obtained in the basin, together with a large water area, whilst the contents of the basin are thoroughly drawn out. Great care is necessary, not only to select a thoroughly good form of closet embodying this principle, but also to ensure that it is properly fixed for use.

CHAPTER XIV.

FLUSHING AND WATER-WASTE PREVENTING CISTERNS.

FLUSHING AND WATER-WASTE PREVENTING CISTERNS:— Water for drainage purposes to be distinct from domestic supply—Object of water-waste preventing cisterns—Quantity of water necessary for closet flushing purposes—The requirements of a good water-waste preventer—Valve closets to be provided with an after-flush water-waste preventer—Construction of a supply valve and bellows regulator—Sizes of supply valves and pipes for valve closets—Flushing cisterns for urinals.

THE water necessary for sanitary purposes must be kept entirely distinct from that required for general domestic use. The supply to closets, slop sinks, &c., should therefore never be taken direct from a service pipe or cistern which is used for any other purpose. In addition, the overflow from all storage or other cisterns must discharge into the open air, away from any soil or ventilating pipe in connection with the drains, so that the water supply may not be contaminated either directly or indirectly with sewage or any impure gases arising from such matters.

Most water companies insist on what is known as a water-waste preventing cistern being fixed to all closets or other sanitary appliances requiring a water-flushing arrangement. This is to ensure an economical expenditure of water, and consists of a small cistern capable of containing and supplying a limited quantity of water sufficient for one flushing operation. Where water-waste preventing cisterns are provided, the sanitary requirements of the case are also at the same time complied with, by the water supply to the fitment being at this point disconnected from the general domestic supply.

Flushing cisterns, or water-waste preventers, vary in capacity from one gallon upwards; but, as a rule, water companies will not permit the use of a water-closet waste-preventing cistern in which the consumption of water is greater than two gallons for each flush, and in some cases will only allow certain specified and approved types of flushing cistern or water-waste preventer to be used.

Generally, it may be said that two gallons of water is quite inadequate for the proper water-carriage or removal of fæcal matters from the soil pipes and drains into the public sewer whilst the minimum size of the soil pipe and drain is fixed at 4 inches diameter. The consequence is that the greater portion of the fæces remain within the house drains, in a more or less advanced state of decomposition, until such time as they may be removed by successive volumes of flowing sewage or water. The Sanitary Institute, from the results of some experiments recently carried out, recommended that a minimum of three gallons, and a maximum of three and a half gallons, should be allowed for water-closet flushing purposes. Where circumstances permit, a three-gallon flush should be given to all water-closets, and a two-gallon flush to housemaids' slop sinks.

Water-waste preventers, or flushing cisterns to closets and slop sinks, are generally so arranged that on pulling or lifting a handle the water in the cistern is rapidly discharged by siphonic action into the basin below, the fitment being thoroughly cleansed and the contents removed by the powerful flush thus obtained.

A water-waste preventer of a good type should flush properly when the handle is gently pulled, and the discharge should be rapid and certain, whether the handle is held during the whole period of discharge or not. For this reason it should be valveless, the cistern being emptied by siphonic action. The angles should be rounded, and the sides slightly sloped, so that it may not be fractured by frost. The supply pipe from the water-waste preventer to a closet of the wash-down form, should be of strong $1\frac{1}{2}$-inch diameter lead pipe, the upper end being soldered to the brass union outlet of the waste preventer. The lower end of the pipe should be secured to the flushing arm of the closet by means of a strong india-rubber cone properly bound with strong copper wire, as shown in Fig. 58. A single flush given to slop sinks and closets of the wash-down type should be all that is necessary to remove the contents of the basin and recharge the trap with perfectly clean water.

With regard to *valve-closets*, the ordinary single-flush water-waste preventer is unsuitable, for if the valve be unwittingly kept open too long, the whole of the water in the waste-preventer cistern is discharged, and none remains to fill the closet basin when the valve is closed.

In cases where a water-waste preventer must be used with a valve closet, it should be of the description known as an *after-flush water-waste preventing cistern*. This is designed to give both a *flush* and distinct *after-flush* to the basin, and must be so arranged that if the handle of the closet is held up and the valve under the basin kept open until the first flush ceases, the second or after-flush will not flow into the basin until the handle is released and the valve closed, so that the basin may be left with a proper depth of water therein.

The method usually adopted for flushing valve-closets in places where water-waste preventing cisterns are not compulsory, is to fix a small cistern (disconnected from the remainder of the domestic water supply) overhead, and supply the closet therefrom by means of a *supply valve and bellows regulator* fixed under the seat. A sketch of such an arrangement is given in Fig. 60; but the closet basin and trap are omitted for the sake of clearness. On lifting the handle, the closet-valve and the water-supply valve V are both opened, and water is admitted to the basin so long as the handle remains raised. On releasing the handle the closet-valve is immediately closed, whilst at the same time the water-supply valve is also being gradually closed by means of the weighted lever L attached to the supply valve and the bellows regulator. The rate of descent of the lever L may be regulated by means of the tap T, which allows the air within the air-cylinder or bellows regulator to escape at any desired speed. Accordingly, as the air is permitted to escape either slowly or quickly from the regulator, so will the lever descend, the water-supply

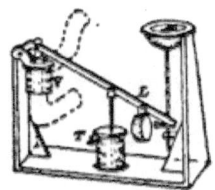

FIG. 60.

valve being closed at a corresponding rate, and the volume of water delivered into the basin after the closet-valve is closed will be proportionately determined.

In another type of valve, designed to govern the supply of water to valve closets, the bellows regulator is omitted and the regulation of the water supply effected by mechanism forming a component part of the valve itself. They are commonly known as "regulator closet valves," or "waste-preventing closet valves," and in general appearance are similar to Fig. 60, but without the bellows regulator shown at T. To meet the requirements of some water companies the waste-preventing valves are so arranged that should the closet handle be carelessly left raised, the supply valve closes directly the requisite amount of water flush has been given, whilst the after-flush is not given until the handle is lowered and the closet-valve closed.

The size of the water-supply pipe and valve will depend upon the head of water available. The following sizes will afford a good flush where the cistern is fixed above the closet at the heights mentioned, viz.:—

Size of Supply Valve and Pipe for Valve Closets.		
Head of water available, or height of cistern above closet.	Internal diameter of supply pipe to valve.	Internal diameter of supply valve.
2 feet.	$2\frac{1}{2}$ inches.	2 inches.
4 ,,	2 ,,	$1\frac{1}{2}$,,
8 ,,	$1\frac{1}{2}$,,	$1\frac{1}{2}$,,
12 ,,	$1\frac{1}{4}$,,	$1\frac{1}{4}$,,

All cisterns should be provided with an overflow pipe discharging into the open air. The sizes and descriptions of flushing cisterns required for flushing branch and main drains have already been mentioned. For urinals where a constant flush cannot be given, a one-gallon automatic flushing cistern discharging at frequent intervals should be provided.

CHAPTER XV.

WATER-CLOSET CONNECTIONS.

WATER-CLOSET CONNECTIONS:—Importance of proper connection between closet and soil pipe—Joint for lead pipe—Connection of lead to iron soil pipes, or lead to stoneware pipes—Method of jointing earthenware with lead, iron, or stoneware pipes—Connection of iron to stoneware soil pipes—Joint for iron to iron pipes—Joints of anti-siphonage pipes—Special form of joint made with an iron collar—Soldering lead pipes directly to earthenware—Connection between closet and soil pipe to be accessible—Description of joint to be avoided—Outlets of closets to be of P form.

IN addition to the provision of a satisfactory form of water-closet, it must be remembered that the sanitary efficiency of the fitment will depend in a great measure on the proper fixing of it, and on the character of the joint between the closet trap and the soil pipe. The consideration of the form of closet connection to the soil pipe may appear to be of minor importance, and unfortunately is too often neglected in practice; but it is essential that the greatest care shall be exercised in securing a permanently satisfactory joint at this point. The drains and soil pipe may be properly constructed in every respect, the closet appliance may be the best of its kind and well adapted for the purpose for which it is intended, yet, if the water-closet be fixed in a slovenly manner or imperfectly connected, the use of it may be absolutely dangerous to health.

It is therefore most necessary that the connection made between the closet trap and soil pipe, on the drain side of the trap, shall be perfectly air and water-tight, and altogether above suspicion; otherwise there is great risk of the vitiated air from the drains escaping through the defective joint into the building. When the connection is between materials of the same description, as lead and lead, stoneware and stoneware, &c., the joint is, or should be, easily and securely made; but when it is required to connect two different materials, as stoneware to lead, stoneware to iron, or *vice versâ*, lead to stoneware, iron to stoneware, the formation of a proper joint at this point is rendered somewhat difficult.

Where valve closets are used, the trap under the closet should invariably be of lead, the connection to the lead soil pipe being made with the usual wiped solder joint. The connection between a lead trap and an iron soil pipe must be effected by means of a strong brass ferrule, secured to the outlet of the lead trap with a wiped joint, as shown in Fig. 61. The lead is dressed over the other end of the ferrule and inserted in the socket of the iron soil pipe, the joint being then run with lead and caulked in the usual manner. Should it be necessary to connect the lead trap with a stoneware drain or soil pipe, the joint is made in exactly the same way as just described (see Fig. 61), except that neat Portland cement is substituted for the molten lead.

When the outlet of the trap is of earthenware and the soil pipe of lead, the joint should be made with a brass socket or thimble, as shown in Fig. 58. The thimble is soldered to the soil pipe, the outlet of the trap being then inserted into the brass socket, and the joint made with neat Portland cement.

Where the outlet of the trap is of earthenware and the soil pipe of iron or stoneware, the end of the trap

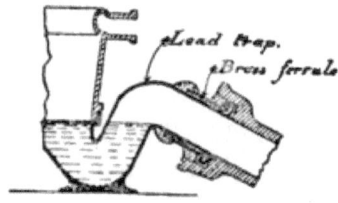

Fig. 61.

must be inserted into a socket on the soil pipe itself, and the joint made good with neat cement, as shown in Fig. 62.

In cases where the trap outlet is of iron and the soil pipe of stoneware, the end of the trap must be placed in the socket of the soil pipe and the joint made with cement. When both are of iron the joint would, of course, be run with lead and caulked.

All the above-mentioned joints, when properly made as described, will be found to comply with the requirements of the bye-laws of the London County Council.

The joint between the anti-siphonage pipe and the

closet trap requires to be just as carefully made as that between the closet trap and soil pipe, and should be carried out in the same manner in all its details. Fig. 62 shows the connection between a stoneware closet trap

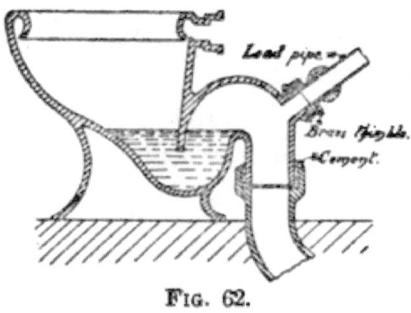

FIG. 62.

and a lead anti-siphonage pipe, the joint being made by means of a brass thimble soldered to the lead pipe.

Owing to the difficulty of making a satisfactory joint with Portland cement between the earthenware

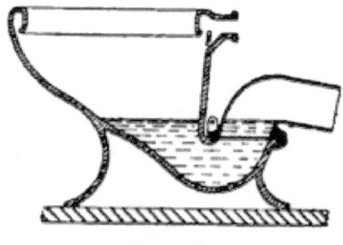

FIG. 63.

trap of the ordinary flush-down closet and an iron or lead soil pipe, various special forms of joint have been devised, of which a few types will be briefly mentioned here.

Fig. 63 is a sketch of a well-known wash-down closet, in which the connection between the stoneware and lead is so arranged as to be below the water level of the closet trap, the joint being secured by means of an iron collar, which is capable of being tightened up with small screw bolts. Being thus permanently submerged, any defect at this connection would immediately become apparent. The lead outlet can accordingly be connected to a lead soil pipe by means of the usual wiped joint.

Another type of joint recently introduced consists in soldering the lead pipe directly to earthenware by

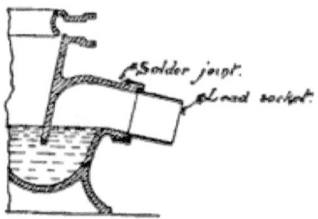

FIG. 64.

means of a special flux or soldering material, so that a thorough incorporation of the two materials is effected at the junction, and a satisfactory air and water-tight joint obtained. Fig. 64 shows a well-known form of this joint, in which the closet is sent from the manufacturers with a lead socket securely attached to the outlet of the earthenware trap with a patent soldered joint. Each of these joints is tested to withstand a pressure of 45 feet head of water before being sent out. A wiped joint readily secures the socket to a lead soil pipe or branch. In another patent joint of similar character, the outlet of the earthenware trap is coated with some material which admits of being readily tinned, so that when being fixed, a wiped lead joint may be made *directly* between the earthenware of the closet and the lead soil pipe. It is stated that when joints are properly made in this manner, if subjected to great hydraulic pressure the ware itself would give way rather than the joint.

The connection between the soil pipe and the trap of any wash-down form of closet should always be accessible for examination. Closets having the outlet of the trap formed as shown in Fig. 57 should not be used for this reason. The efficacy of such a joint entirely depends on the perfect bedding and jointing of the closet base upon the end of the soil pipe, and it is very seldom that a thoroughly air-tight joint can be formed in this manner. On the least disturbance of the closet fitment the joint may be broken, and the damage remain unseen and undiscovered until serious consequences to health have been caused by the free entrance of vitiated air from the soil pipe into the apartment through the defective connection.

The outlets of all closet traps should preferably be of the P form, as shown in Figs. 58, 61, 63, &c. Where it is necessary to use the S form of outlet, the joints between the soil pipe and the outlet should be *above* the floor-level, as shown in Fig. 62, so that the whole of the joint may be visible and accessible at all times.

CHAPTER XVI.

BATHS.

BATHS:—Their hygienic importance—Descriptions of baths—Essentials of a sanitary form of bath—Slate baths—Porcelain baths—Marble baths—Zinc baths—Enamelled-iron baths—Copper baths—Average size and weight of plunge baths—Objections to concealed standing overflow and waste—Good arrangement for bath overflow and waste—Alternative arrangement for bath overflow—Water supply to discharge *above* the overflow level of bath—Baths to be unenclosed—Cradling for baths.

FREQUENT bathing of the body is often looked upon as a luxury, instead of constituting a material factor in the maintenance of personal health; but of late years its hygienic importance has been more fully realised, with the result that public baths and bathing-places are now established in all our crowded cities.

A simple and sanitary form of bath is, or should be, a necessity in every house. In this connection it may be mentioned that some years ago the governing body of the city of Baltimore passed a law making it compulsory to provide a bath-room in *all* domestic dwellings which should be erected in that city, from the costly mansion down to the small cottage of the artisan.

The general arrangement and details of construction

of baths depend in a great measure on the particular uses for which they are required. For medical and other purposes, special forms, such as "needle" baths, "sitz" baths, "vapour" baths, &c., are frequently used; but for ordinary domestic requirements the *plunge* or *slipper* bath is almost universally adopted, though sometimes it may be fixed in combination with a water spray apparatus, so as to give a douche, spray, wave, or shower bath when desired.

A well-designed and properly constructed *plunge-bath* should fulfil the following conditions in order to afford an adequate measure of sanitary efficiency, viz.:—

1. The material of which the bath is made should be perfectly smooth and impermeable, and contain no angles or corners for the retention of dirt or soapy matters.

2. The waste pipe must be entirely disconnected from the drain, and securely trapped close to the outlet of the bath, whilst the trap should be of siphon form with a good water seal, and thoroughly self-cleansing. An anti-siphonage pipe should also be fixed near the top of the trap.

3. Every portion of the bath and its fittings should be visible and easily cleaned, no portion of the overflow and waste pipes situated on the *bath* side of the trap being concealed or inaccessible at any time.

4. The water supply to the bath should be completely disconnected from the waste and overflow pipes.

5. The waste pipe should be as large as practicable (about 1½ inches or 2 inches diameter), so as to afford a "quick" waste for flushing the drains. The overflow pipe should be large enough to carry off all the water discharged by the supply valves when turned on full bore.

6. A properly constructed safe (discharging into the open air) should be provided.

Plunge baths are usually made of slate, porcelain, marble, zinc, cast iron, or copper. An absorbent material like wood is, of course, quite unsuitable.

Slate baths generally consist of planed and enamelled slate slabs bolted together, the joints being made with red lead. Such baths possess too many objectionable angles and corners to be satisfactory.

Porcelain or fireclay, when well glazed and without flaw, provides a thoroughly impervious material with a perfectly smooth and even surface. Owing to the thickness of material necessary for strength, a great deal of heat is absorbed from the hot water when a warm bath is required. They are expensive in first cost, extremely heavy, but when fixed are most cleanly and lasting in wear. They should only be used where an ample supply of hot water is available, so that the bath may be well warmed before use. In cold weather the temperature should be raised gradually, in order to avoid any risk of fracture.

Marble baths are cut out of the solid block, and afterwards polished. For ordinary purposes they are prohibitive in price, and have no sanitary advantages over those of well-glazed porcelain. Like porcelain, they are very heavy, and absorb a great amount of heat from any warm water contained therein.

Zinc baths are sometimes fixed where economy in first cost is necessary; but they are unsuitable for general use, as they soon become worn out.

Enamelled cast iron is the material most commonly used for baths on account of its comparative cheapness. So long as the enamel remains intact, a smooth impervious surface is obtained; but should it become cracked or chipped, the exposed surface of the iron rapidly oxidises, the bath being perforated by the rust, and so rendered unserviceable. The cheaper descriptions are either painted with enamel paint or japanned; but where cast-iron baths are adopted they should be "porcelain enamelled," with a good smooth surface and free from all cracks or other flaws.

Sheet copper forms an excellent material for baths, and is very durable. It is not liable to rust if the surface of the metal is exposed through any wear or scratching of the enamel. Being also of comparatively thin substance, it absorbs but little heat from any warm water contained by it. When this material is used, the bath should be constructed of stout hammered copper weighing not less than 32 ounces per foot super., with good lap, welded and brazed seams, the ends and bottom being rounded, and the whole well tinned and enamelled.

Plunge baths may be obtained with tapering or parallel sides, and of sizes varying from 5 feet to 6 feet in length. The average internal dimensions of what is known as a "full-size" bath are as follows:—

    Length, 5 feet 6 inches.
    Breadth at head, 2 feet 1 inch.
    Breadth at foot, 1 foot 8 inches.
    Depth, 1 foot 10 inches.

The comparative weights of a full-size plunge bath of different materials is given in the following table:—

Average Weight of a Full-Size Plunge Bath.	
Description of Material.	Weight.
Sheet copper	76 lbs.
Enamelled cast iron	300 ,,
Slate	500 ,,
Porcelain	500 ,,
Marble	600 ,,

Whilst with but few exceptions nearly all baths are now made with rounded angles and corners, yet the great majority of them, as at present constructed and fixed, do not provide that degree of sanitary efficiency that might reasonably be expected. The greatest defect is usually found in the arrangement and method of fixing the waste and overflow pipes.

One of the most popular, and at the same time —from a sanitary standpoint—objectionable arrangements, is that known as a "secret" or "concealed" standing overflow and waste, the general form of which is shown in Fig. 65. The bath waste is trapped as shown at T, the inlet of the trap being tapered in order to provide a seating for the standing overflow pipe, which is capable of being raised or lowered at will by means of a small knob or pull. Should the standing

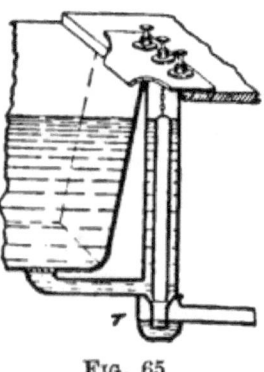

FIG. 65.

overflow pipe be raised, the water is free to escape through the trap and waste pipe; but when resting upon its seat the contents are retained until the level of the top of the standing overflow is reached, the surplus water then passing down the overflow pipe and so through the trap below. The standing overflow being thus made to act as a plug to the outlet of the bath, it will be observed that on every occasion when the bath is used the annular space between the overflow pipe and the tube containing it must, of necessity, fill with water to the same level as in the bath. After continued use, the surfaces of the annular space—together with that portion of the waste pipe between the plug and the bath grating—become fouled with soap, grease, dirt, &c., the whole being either out of sight and inaccessible, or so inconveniently arranged that they cannot be properly cleaned. The air of the bath-room is consequently liable to become vitiated by the gases generated and given off from the decomposing organic matters. At the same time, whenever the fitment is used, the clean water entering the bath becomes more or less contaminated by the decomposing matters adhering to the soiled surfaces of the concealed overflow chamber and the waste pipe. It may be readily imagined that in the case of certain contagious diseases the indiscriminate use of a bath of this description might be attended with serious consequences. It should also be noticed that the dip form of trap shown in Fig. 65 is not a thoroughly self-cleansing type of trap.

A simple and good arrangement is shown in Figs. 66 and 66A. A self-cleansing siphon trap is fixed immediately under the bath outlet; the bath plug takes the form of a standing overflow placed in a small open recess at the foot of the bath, and can be raised or lowered by means of a small lever or knob. The combined plug and overflow may be instantly unhooked at the top and thoroughly cleansed when necessary. By this means every portion of the bath and its fittings is so designed as to be visible and accessible at all times, whilst the overflow pipe, being placed in the small recess, occasions no inconvenience to the bather.

In some districts the use of a standing overflow to the bath—whether visible or concealed—is not per-

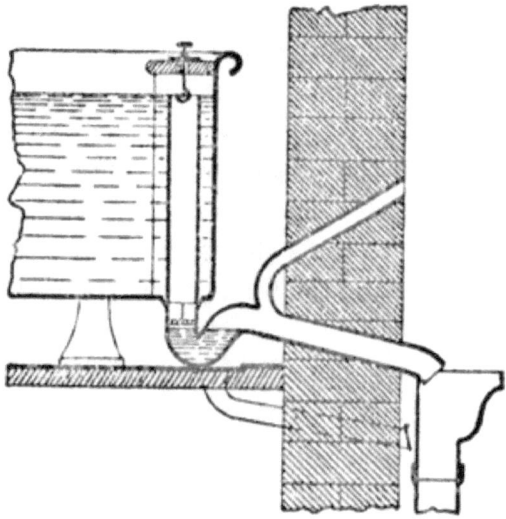

Fig. 66.

mitted by the regulations of the water company. Under such circumstances, it is generally compulsory that the bath overflow shall not in any way be connected or discharge into the waste pipe from the bath, but shall be carried through an external wall as a warning pipe and discharge in some conspicuous position, in order to prevent any waste of water taking place through the overflow without being noticed. The outer end of the overflow should be provided with a brass or copper hinged flap valve, whilst the outlet from the bath may be arranged with a movable grating or other device, so that the whole length of the pipe may be easily cleaned. As all baths should stand upon a properly constructed safe or tray, the overflow pipe, instead of being carried through the wall, may discharge over the mouth of the waste pipe from the safe, the outer end of the safe

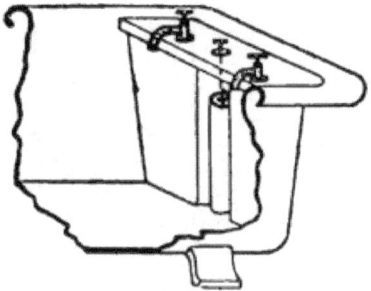

FIG. 66A.

waste being fitted with the usual flap valve. Fig. 67 is a sketch showing the method of fixing a bath so as to comply with these conditions. The bath outlet is provided with a solid pull-up waste in a small open recess; but an india-rubber or brass plug with chain may be substituted if desired. The lead safe under the bath is constructed in the same manner as already described for water-closets. If the bath overflow grating cannot be removed for cleaning purposes, a brass screw cap might be provided to the overflow pipe at A (see Fig. 67).

The hot and cold water supply should discharge *over* the edge of the bath, or, at least, *above* the overflow level of the bath. It may be observed that, according to the London Metropolitan Water Act of 1871, it is

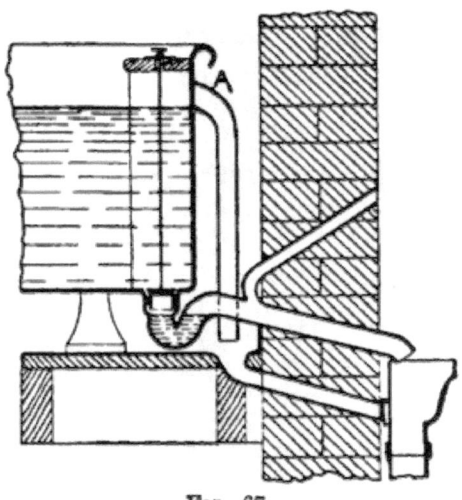

Fig. 67.

required "that the outlet shall be distinct from, and unconnected with, the inlet, or inlets; and the inlet, or inlets, must be placed above the highest water level of the bath." By this means any possible contamination of the water supply is reduced to a minimum. At one time it was a common practice to admit hot and cold water into the bath through the same orifice as that used for the outlet of the waste pipe. Such an insanitary arrangement should not be permitted, for particles of soap, grease and other impurities are from time to time deposited on the interior surface of the pipe common to the waste and water supply, a great portion of which is returned to the bath on each occasion that it is used.

On hygienic grounds it is desirable that the bath should be quite unenclosed, with sufficient space between it and the walls to allow of every part being

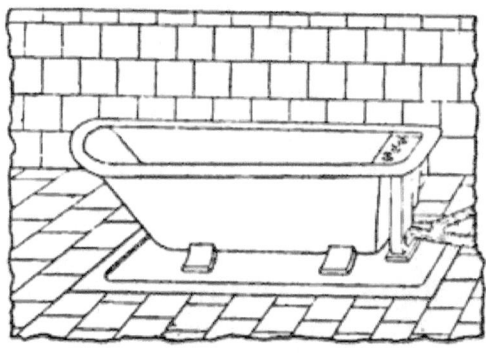

Fig. 68.

readily cleaned (see Fig. 68). The upper part of the bath should be formed with a rolled edge. The floor of the bath-room should be laid with mosaic, tiles, or other non-porous material, the bath standing within a properly constructed safe, which may be made of slate, marble, glazed earthenware, or tiles. The safe should be provided with a $1\frac{1}{4}$-inch diameter waste pipe discharging into the open air, the outer end being provided with a flap valve. Where no bath enclosure is fixed the exposed supply pipes, waste pipe, &c., may be of polished brass.

If the bath is made of thin sheet metal, it must be supported with a wooden framing or "cradling," so as to prevent the sides and bottom being forced out of

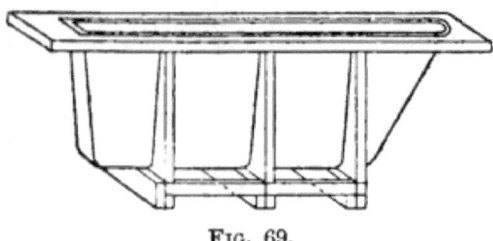

Fig. 69.

shape when in use. It is then necessary to fix a wooden enclosure to the bath for the sake of appearances, and under such circumstances the safe should cover the whole of the floor surface within the enclosure. Fig. 69 is a sketch of the cradling as usually fixed.

CHAPTER XVII.

HOUSEMAIDS' SLOP SINKS.

Housemaids' Slop Sinks:—Combined closet and slop sink—Housemaids' washing-up sink—Combined slop closet, and wash-up sink—Size of soil pipes for slop closets—Slop sinks to be in well-lighted and ventilated lobbies.

For a small dwelling the simplest method of disposing of fouled chamber liquids is to arrange the sanitary fitments so that a water-closet—preferably one used by the servants—is available for this purpose. In such a case it is desirable that the combined closet and slop sink should be of the wash-down pedestal form, and fitted with a circular-fronted white-glazed earthenware

Fig. 70.

slop top under the hinged seat of the closet. Fig. 70 is a sketch of a slop top suitable for this purpose. A housemaids' washing-up sink must also be provided in some convenient place. The sink may be of glazed stoneware or enamelled cast iron, supported on cast-iron brackets, and the whole of the appliance, including the waste and overflow pipe, should be visible and readily accessible for cleaning purposes.

Instead of the usual brass washer and plug to the outlet of the waste and a concealed overflow pipe, it is better to provide a standing waste and overflow, which may be placed in a small open recess, as shown

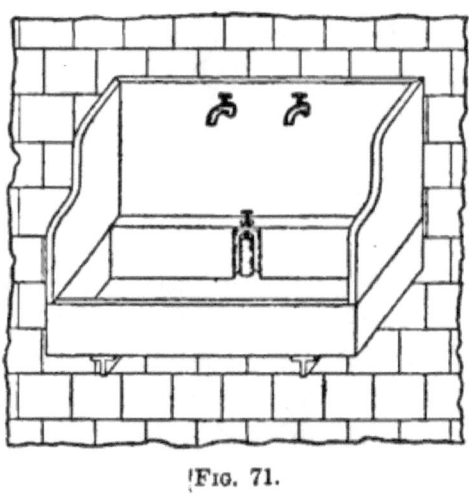

FIG. 71.

in Fig. 71. A siphon trap must be fixed immediately under the sink outlet, the waste pipe discharging outside over the hopper head of a vertical waste, which should again discharge over a trapped gully. Both hot and cold water supply are usually laid on to the sink.

In large establishments, where separate slop closets and sinks are necessary, the slop sink and wash-up sink may be combined, as shown in Fig. 72. The slop sink must be fixed under precisely the same conditions

as required for a water-closet. The whole should be
thoroughly self-cleansing, the basin being provided with

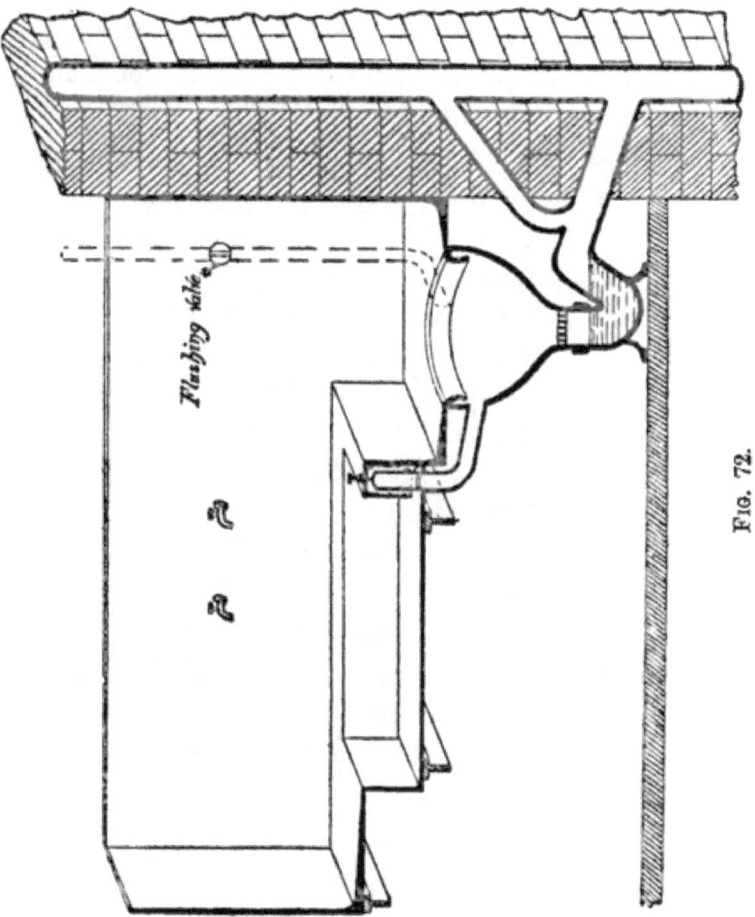

Fig. 72.

a flushing rim and small flushing cistern or water-
waste preventer. A loose grating should be fitted at the
bottom of the basin, in order to retain pieces of soap,

brushes, cloths, or other foreign articles, which might at any time be carelessly thrown in with the slops.

The soil pipe and trap to a slop sink need not be larger than $2\frac{1}{2}$ inches or 3 inches diameter. It must be connected to the drain, and carried up full bore above the eaves as a ventilating pipe, in the same way as described for the soil pipe of a water-closet. An antisiphonage pipe should also be fixed near the top of the trap, as shown in Fig. 72. The wash-up sink should be provided with a standing waste and overflow pipe in an open recess at the side, the waste pipe therefrom discharging into the slop sink just *below* the flushing rim.

Housemaids' slop sinks should be fixed in well-lighted and ventilated lobbies, not in some dark, inconvenient, out-of-the-way corner, as is so frequently the case. The floor should be of concrete, tiles, mosaic, or other similar materials, the portions of the walls contiguous to the slop sink being protected with enamelled slate or glazed tiles to a height of at least 3 feet. The hot and cold water supply must not be placed over the slop sink, as the water drawn from taps in such a position might possibly become fouled with splashings from the slops and also by dirty cloths being temporarily placed upon them. The water supply should be arranged to discharge over the wash-up sink, as shown in Fig. 72.

Both the slop closet and the wash-up sink are best fixed without any wooden enclosure, so that every part may be visible and readily cleaned.

CHAPTER XVIII.

LAVATORIES.

LAVATORIES:—Requirements of a good lavatory basin—Fixed basins with plain rim—"Flushing-rim" basins—"Tip-up" basins—Lavatory tops—Defects of lavatory basins with concealed overflow—Objections to basins fitted with concealed standing overflow and waste—Simple form of standing waste and overflow—Arrangement of lavatory fittings when a standing waste and overflow is not permitted—Water supply to lavatory basins—Lavatory wastes—Anti-siphonage pipe to be provided.

THESE consist of one or more wash-hand basins conveniently arranged for ablution purposes, and having a supply of hot and cold water laid on to them. As a rule, the basins are either *fixed* or *pivoted*, the latter description being also known as "tip-up" lavatory basins. The basins are usually made of glazed porcelain, or, when liable to be subject to rough treatment, of enamelled cast iron. They should be circular or oval on plan, with an entire absence of sharp angles or corners within the bowl, and the whole, as far as possible, self-cleansing.

Fixed basins generally have a plain rim, as Fig. 73; but they can also be obtained with a flushing rim, as shown in Fig. 74, in which case the water supply may be turned on for a few seconds in order to thoroughly cleanse and flush the sides of the basin before use.

"Tip-up" or "lift-up" basins were at one time very popular for lavatory fitments, and are still much used. They consist of a bowl pivoted upon a receiver, into which the contents of the basin are discharged when

Fig. 73.

turned upon its trunnions (see Fig. 75). Any danger of fracture when the basin is carelessly swung is avoided by means of an indiarubber buffer. The overflow weir or pipe from the basin discharges into the receiver below. Tip-up basins do not provide a thoroughly

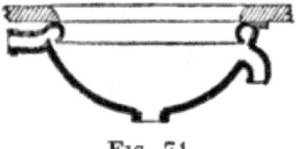

Fig. 74.

satisfactory form of lavatory, for the surface of the receiver and the under side of the basin become fouled with the constant discharge of dirty and soapy water. The bowl is frequently so arranged that it can be lifted out for the purpose of cleansing the receiver; but even if the greatest care be exercised in the frequent cleaning of the fitment, the provision of such a large and unnecessary fouling surface on the basin side of the trap, and which in itself is not self-cleansing, is a serious sanitary objection.

The lavatory top may be of slate, marble, porcelain, or other impervious material. Where marble is used, it should be properly bedded in plaster of Paris, and not in oil putty, as the oil is liable to penetrate the marble, causing unsightly stains which cannot afterwards be removed.

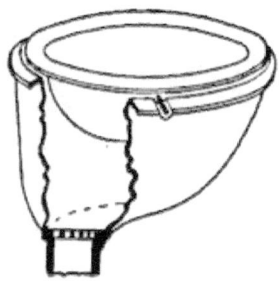

FIG. 75.

The lavatory fittings are generally placed within a wooden enclosure for appearance sake; but it is better to support the fitment on brackets, so that everything may be exposed to view.

A good lavatory requisite should fulfil, as far as possible, the conditions already mentioned for the proper construction of an efficient plunge bath, and which need not be again repeated. Perhaps the most serious sanitary defect to be found in the average lavatory fitment with a fixed basin is the almost universal use of a concealed and inaccessible overflow pipe, together with a comparatively large soiling surface between the outlet of the basin and the plug.

Fig. 76 is typical of such an arrangement, having a solid waste plug with an inaccessible overflow arm. Instead of the trap being placed directly under the

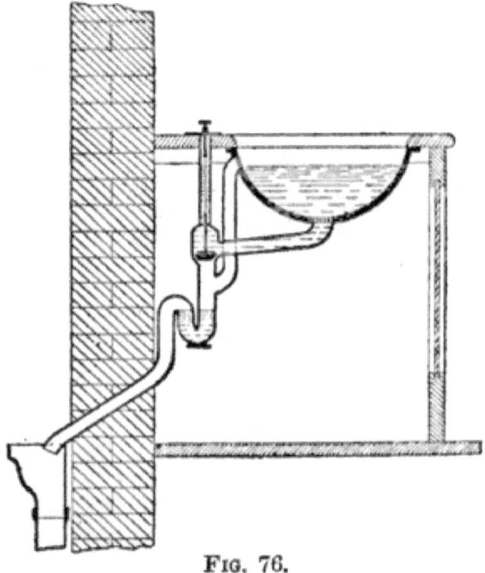

Fig. 76.

outlet of the basin, it will be seen that there is a very large fouling surface between the basin outlet and the plunger, which, in the course of time, gradually becomes furred with greasy and soapy deposits. The overflow arm also becomes fouled in the same manner, and is ordinarily so fixed that its interior cannot either be readily inspected or cleaned. These defects are also

LAVATORIES. 139

apparent in the common form of lavatory basin, a sketch of which is shown in Fig. 77.

Lavatory basins provided with a concealed standing overflow and waste, as shown in Fig. 78, should be avoided. The objections to this form of waste and overflow as applied to baths have already been indicated,

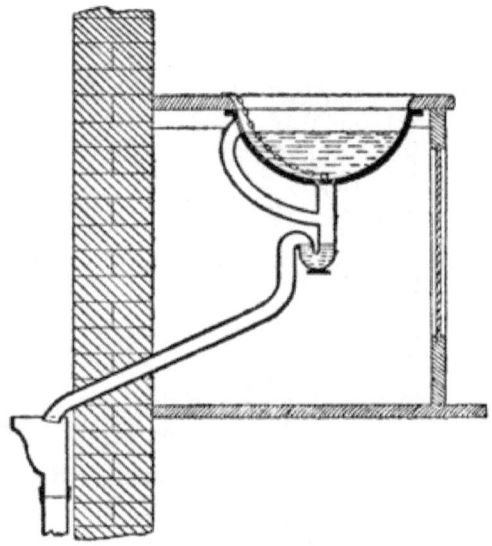

Fig. 77.

(see Fig. 65), and the same remarks apply with equal force to any similar arrangement for lavatory fittings. In Fig. 78 a siphon trap is shown under the standing overflow, in place of the dip form seen in Fig. 65.

A simple form of standing waste and overflow pipe is shown in Fig. 79. This is fixed in an open recess at the back of the basin, with a siphon trap immediately

140   SANITARY HOUSE DRAINAGE.

under. The waste is opened or closed by raising or lowering the stand pipe, a half turn keeping it open

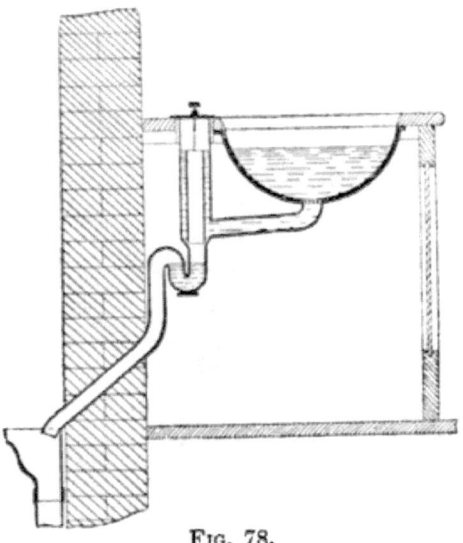

Fig. 78.

when the basin is required to be emptied. Every part of the waste and overflow on the basin side of the trap is visible and easily cleaned.

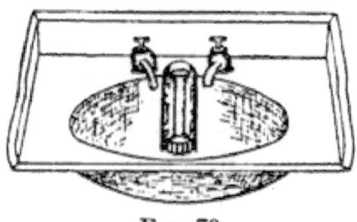

Fig. 79.

Another form of exposed standing waste and overflow is shown in Fig. 80, the details of which are similar

LAVATORIES. 141

to those described for a standing waste and overflow for baths (see Figs. 66 and 66A).

In some districts the water company will not permit the use of any description of overflow pipe which is arranged to discharge into the lavatory waste pipe, but insist upon the overflow being carried through an outer wall so as to discharge in some prominent position.

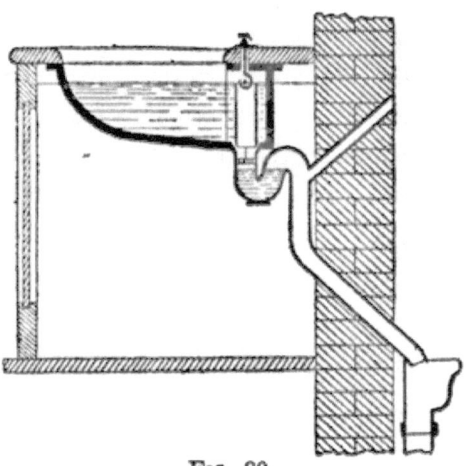

FIG. 80.

Under such circumstances no form of standing waste and overflow can be adopted; but the basin overflow must either discharge directly into the open air, or be arranged to discharge into a safe or tray fixed directly under the basin in a manner similar to that already described for baths fixed under the same conditions (see Fig. 67).

The water supply should discharge *above* the highest

water level of the basin, whilst the provision of self-closing taps is also enforced by some water companies. Lavatory basins which admit the water supply through the waste pipe should be avoided altogether.

The waste pipe should not be less than $1\frac{1}{4}$ inch or $1\frac{1}{2}$ inch diameter, in order that a rapid discharge may be obtained, and also to assist the flushing of the drains. When a perforated grating is fixed to the basin outlet, it should be large enough to allow of the waste pipe and trap running full bore when the basin is being emptied. The lavatory waste should discharge outside the building over a trapped gully or the hopper head of a vertical waste, as already described. An anti-siphonage pipe should also be fixed near the top of the siphon trap to the waste, as shown in Fig. 80.

CHAPTER XIX.

URINALS.

URINALS:— Composition of urine — Urinals generally unfavourably situated—Necessity for ample water supply — Objections to the use of closets as urinals—Essentials of a satisfactory urinal—Advantages of basin urinals for private residences—Good general arrangement of basin urinal—Objections to treadle-action flushing appliances—Urinal compartments to be constructed of non-porous materials— Enamelled iron urinal basins — Self-flushing urinal basins—Details respecting urinal divisions—Stall urinals—Stall urinals with semicircular backs — Trough urinals — Automatic flushing tanks for urinals.

AMONGST the various sanitary appliances that may be required for general domestic purposes, this class of convenience is probably the most difficult to arrange so that it may remain permanently hygienic and inodorous when in constant use. This is due in a great measure to the chemical composition of urine itself.

Urine is a compound of urea, uric acid and other organic and inorganic matters in combination with water. The urea present in urine rapidly decomposes, especially if any slight degree of heat be present. During this process of decomposition large quantities of ammonia are given off, the pungent and unpleasant odour from which frequently permeates the whole of the room in which the urinal is situated. Uric acid is but slightly soluble, and has a tendency to adhere to any surface with which it comes in contact, where it eventually decomposes if not immediately removed.

Urinals for domestic purposes are usually situated in some confined and insufficiently ventilated apartment, having a much warmer atmosphere than the external air. The most favourable surroundings are accordingly afforded for the rapid decomposition of the urine unless it is at once effectually carried away. No description of fitment can, therefore, be considered as likely to be permanently satisfactory unless an ample supply of water with a good flush is provided, so as to quickly and thoroughly remove all traces of urine immediately after its deposition.

Sometimes a water-closet of the wash-down pedestal type with hinged seat, is designed to serve the purpose of a urinal in places where such a convenience would only occasionally be required; but such an arrangement is not suitable for general or constant use. Considered as a urinal fitment, the basin of the water-closet is placed at too low a level for the purpose, and occasions a certain amount of splashing over and around the closet. Any careless droppings of urine are also deposited on the floor, where they are usually allowed to remain and decompose, giving rise to that obnoxious smell at present characteristic of far too many urinal apartments.

To provide an efficient and permanently satisfactory urinal, it is desirable that the following conditions should be complied with as far as possible, viz. :—

1. The apartment in which it is situated should be thoroughly well lighted and ventilated; the floor and walls—or at least those portions immediately contiguous to the urinal—being of some smooth, non-absorbent material.

2. The amount of soiling surface over which the urine may be spread should be as small as possible consistent with convenience, and should be entirely free from any angles or corners which might tend to retain deposits of urine or dirt.

3. Every portion of the soiling surface of the urinal fitment should be self-cleansing, smooth, impervious, and not readily acted upon by uric or other acids.

4. An adequate supply of water should be available for the complete removal of all traces of urine on every occasion that the appliance is used.

Urinals may be broadly divided into what are known as *stall*, *trough* and *basin* urinals.

For private establishments, &c., where it is desirable or necessary that the urinal shall be fixed within the dwelling, it is considered that the most satisfactory appliance is afforded by a good type of urinal basin, with a properly constructed foot-plate or base under. The basin should be of glazed porcelain, with lip and flushing rim, the back being angular or flat, according to the position in which it is to be placed. Fig. 81 is a sketch of a flat-back lipped urinal basin, and Fig. 82 an angular lipped basin, both being provided with flushing rims.

The advantage of a urinal basin over that of a stall or trough urinal is that the discharge of urine is concentrated at one definite point, instead of being spread over a large surface area. In stall urinals the whole of the floor, sides and back of the compartment is in a more or less wet, splashed and uninviting condition. This creates a desire on the part of the user to avoid

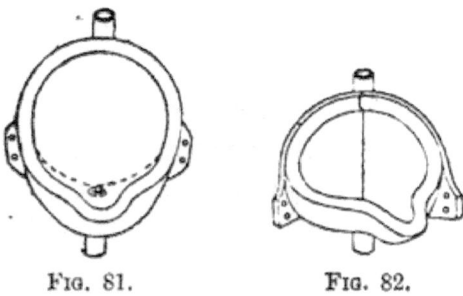

Fig. 81.   Fig. 82.

entering the compartment, so that the discharge is frequently made at some distance from the fitment, the area of the soiled surface being consequently still further enlarged.

Figs. 83 and 84 show the elevation and section of a urinal fitted with a white glazed porcelain flat-back lipped basin, with flushing rim and back outlet. The foot-plate or floor of the urinal compartment is dished to proper falls, so that all drippings may enter the glazed earthenware flushing-rim gully trap. Both the drip gully and urinal basin are trapped, and discharge into a trapped gully outside.

An alternative arrangement of base plate is shown

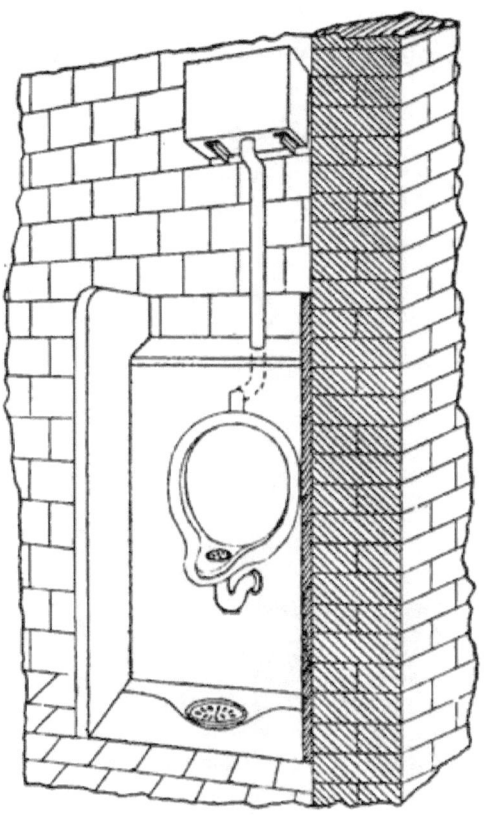

FIG. 83.

in Fig. 85, having a perforated brass sparge pipe and shield at bottom, together with an ordinary trapped gully.

Concerning the water supply, the most satisfactory

results are obtained when it is so arranged that a constant flow of water is given to the urinal basin, together with a frequent automatic flush to the drip gully or sparge pipe at the bottom of the base plate. Many water companies will not, however, permit a

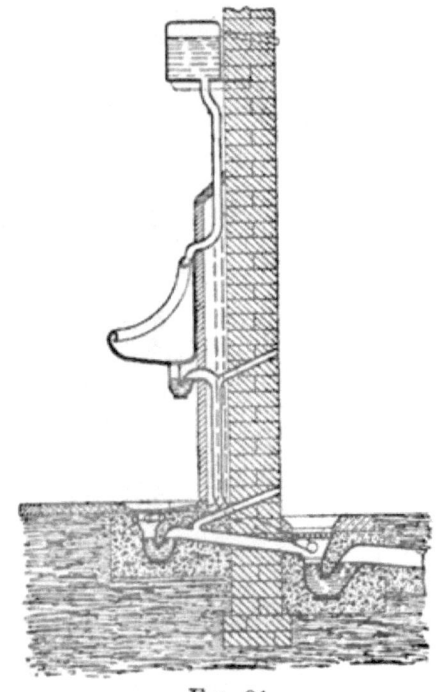

Fig. 84.

constant supply of water to be given, in which case the basin, and also the drip gully, should be connected with an automatic flushing tank discharging at frequent intervals; or arrangements may be made whereby they may be flushed after use by pulling the handle of a water-waste preventing cistern. The proper flushing of the urinal under the latter arrangement is apt to be neglected by the person using the fitment, as the pulling of the handle after use is frequently omitted. Urinals which are flushed by means of a treadle action platform are objectionable, as the apparatus is liable to become fouled and evil smelling.

The back and sides of the urinal compartment may be of marble, glazed porcelain, or enamelled slate; whilst to ensure absolute cleanliness and freedom from

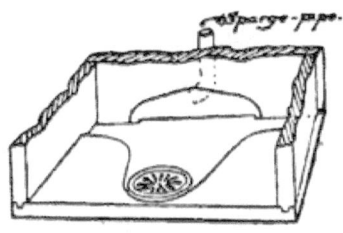

Fig. 85.

smell, it is advisable that the entire compartment be well washed and scrubbed down once a day.

Where subject to rough or unfair usage, the urinal basins are sometimes made of porcelain enamelled iron; but for ordinary purposes they are inferior to glazed porcelain.

What are known as "self-flushing" basin urinals are sometimes fixed. In this type of urinal the discharge of urine into the basin starts a siphonic action water-waste preventer, so that the basin is flushed on every occasion that it is used.

When a range of urinal basins is required, the slate or marble divisions should be spaced 2 feet apart, and supported on cast-iron brackets fixed about 2 feet 3 inches above the floor level, so that the lower portion of the whole range may be easily cleaned from end to end. They should project not more than 15 inches, as such a width will be found quite sufficient for decency,

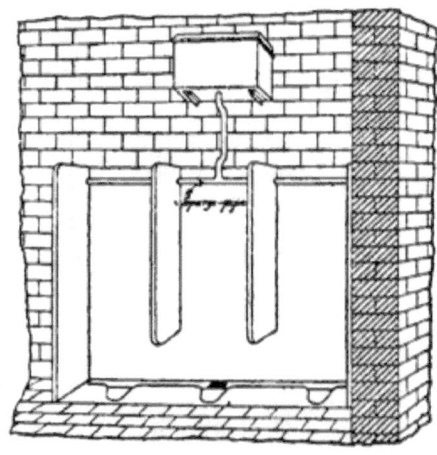

Fig. 86.

whilst ensuring that the user shall stand close to the basin during the discharge.

For large institutions, where the utmost economy of construction and maintenance is a great consideration, "stall" urinals are used. They are placed in an airy enclosure at some distance from the main buildings. Fig. 86 shows a range of urinals of this character, having slate divisions and backs, dished foot-plates, and continuous channel discharging over a trapped gully. A horizontal perforated copper pipe—known as a "sparge" pipe—runs the whole length of the range, the water passing through the perforations in a continuous stream, so as to cleanse the back of the stalls. For sanitary efficiency the soiling surface is much too large, whilst the sharp angles between the back and

FIG. 87.

divisions are objectionable, as they cannot be easily cleaned, and so tend to collect deposits of the salts of urine at those points. The surfaces of the divisions also become greatly splashed with urinary matters, and are totally wanting in self-cleansing action. They are comparatively cheap in first cost, and being generally placed in airy and freely ventilated situations, the surrounding atmosphere exercises a constantly purifying effect upon them, thus remedying in some measure their defective construction.

A much improved form of stall urinal is shown in Fig. 87, and a cross section through the same in Fig. 88. The stalls are formed with a rounded or semicircular back, the sides and floor of each compart-

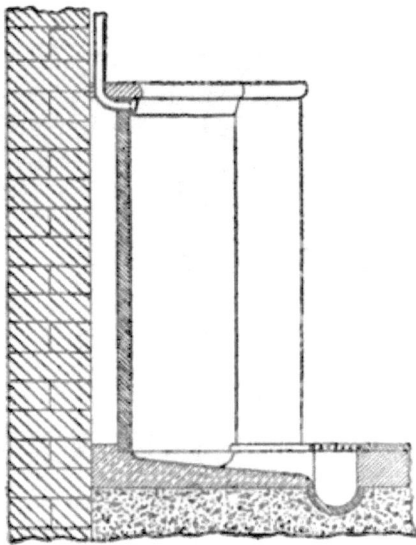

FIG. 88.

ment draining into a channel in front covered with a perforated grating. On taking up the grating the whole length of the channel can be swept and cleaned. Although the soiling surface still remains comparatively large, there are no angles or corners to retain urine or dirt, whilst the whole of the compartment is well flushed with a sparge pipe curved to the shape of the back.

URINALS.

A range of *trough urinals* is shown in Fig. 89, of which Fig. 90 is a section. The urine is discharged into a trough full of water, which is retained within the trough by means of a weir near the outlet. The contents are periodically driven out and the trough refilled with clean water by an automatic flushing tank. As the urine is not immediately removed, the

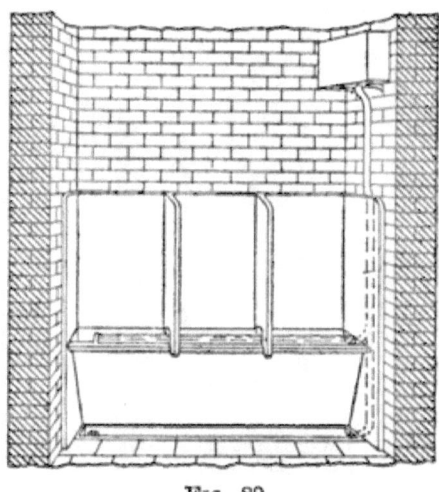

Fig. 89.

surfaces of the trough gradually become coated with deposits from the urine, whilst the front of the range is liable to be fouled from careless usage.

As already stated, it is desirable that the soiling surfaces of all urinals should have a constant stream of water playing upon them, so as to immediately remove the urine and prevent the formation of urinary deposits thereon. Few water companies will, however, permit such an arrangement, and insist upon a water-waste preventing cistern or valve being used. This would meet all sanitary requirements if carried out in its integrity, but unfortunately, where the flushing of the urinal is left to the person using the convenience, it frequently occurs that the user, either through carelessness or haste, omits to discharge the water-waste preventer, and the urinal consequently becomes furred

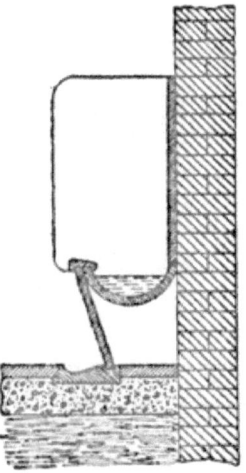

Fig. 90.

with deposits, whilst the atmosphere is tainted with the ammoniacal and other offensive odours arising from the decomposing urine. It is generally more satisfactory to adopt some type of automatic flushing cistern, arranged to discharge at short intervals, so as to ensure the appliance receiving an independent and periodical flush. A one-gallon water-waste preventer will suffice for a single urinal if the discharge is made at frequent intervals.

CHAPTER XX.

TROUGH CLOSETS AND LATRINES.

TROUGH CLOSETS AND LATRINES:—To be detached from inhabited buildings and well ventilated—Trough closets—Latrines—Latrine pans with flushing rims—Abundant water supply necessary—Latrines to be unenclosed—General construction of latrines—Disadvantages of trough closets and latrines—Arrangement of flushing apparatus.

WHEN a large number of closet conveniences are required, it is sometimes found impracticable, under certain circumstances, to provide ranges of ordinary water-closets with an independent water-waste preventer to each. This is more particularly the case where the closet fitments are under no proper supervision, and subject to much rough and unfair usage, and also where economy in first cost and subsequent maintenance is imperative. Where such conditions exist, it is usual to adopt the type of fitment known as "trough closets" or "latrines." They should be invariably placed in a detached outbuilding, well away from any inhabited dwelling, and abundantly ventilated, so that the oxidising influence of large and constantly changing volumes of air may always be present.

A *trough closet* consists of a long channel or trough

of glazed earthenware or enamelled iron, which is partially filled with water for the reception of fæcal deposits, the water being retained within the trough by means of a weir or standing overflow at the outlet. The contents of the trough are discharged from time to time, and the water renewed by hand or an automatic flushing tank. The trough closet may be provided with a continuous front rail for seating purposes, or

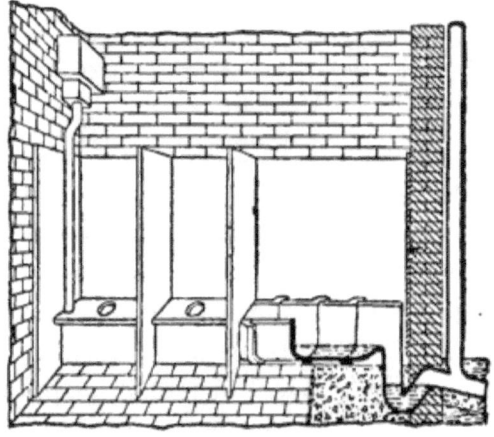

FIG. 91.

hinged seats may be used. Fig. 91 is a sketch of a glazed stoneware open trough closet with hinged seats, weir at outlet, and automatic flushing tank. Every portion of the trough is accessible, and may be readily cleaned by hand with water and a broom.

Latrines consist of a series of basins or pans connected to one common outlet by means of a pipe, and are essentially a modification of the trough closet, having portions of the trough covered at intervals. The contents of the whole latrine range are periodically discharged at the same time, and the water afterwards renewed to all the pans. The latrine pans and pipe connections may be of glazed earthenware or enamelled iron.

Fig. 92 shows a range of three latrine pans with standing overflow and waste, a transverse section

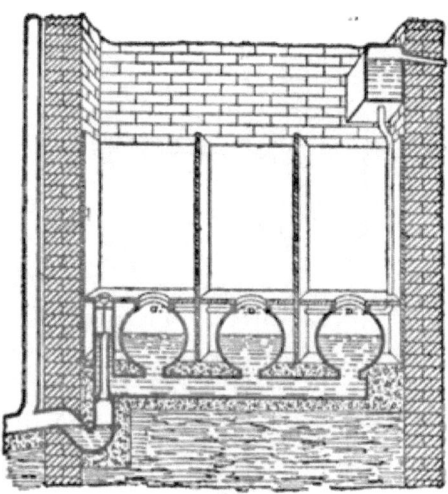

FIG. 92.

through one of the pans being shown in Fig. 93. The range of latrines are emptied, cleaned and refilled by hand at stated intervals.

Latrine pans provided with flushing rims, as shown in Fig. 94, may also be obtained, and if desired, can be arranged so that the whole of the contents are drawn out by siphonic action, the flush, and also the necessary after-

158   SANITARY HOUSE DRAINAGE.

flush for refilling the pans to the required water level, being provided by means of an automatic flushing tank.

Trough closets and latrines should be invariably furnished with an ample supply of water. For trough closets it is usual to provide 10 gallons per seat per day for flushing, cleansing and refilling purposes; whilst for latrines with flushing-rim pans, the daily water supply considered necessary varies from six to eight gallons

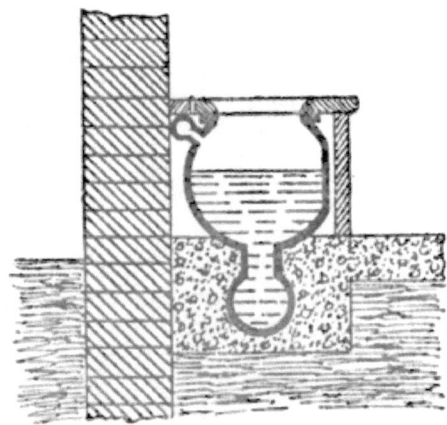

FIG. 93.

per pan. It should be borne in mind that all trough closets and latrines should be trapped at the outlet. (See Figs. 91 and 92.)

It is also desirable that they should, as far as possible, be unenclosed, so that every part may be conveniently reached for cleaning when necessary. Where enclosed, the riser should be of slate or other impermeable material, the intervening space between the latrine and the riser being filled with cement concrete, and floated level with the top of the pan. If wooden seats are used they should be hinged and well painted on the under side. The back wall of the latrine compartments should be lined with slate, or rendered smooth in cement, the divisions being of slate or enamelled cast iron.

Under the most favourable circumstances, trough

FIG. 94.

closets and latrines, from a hygienic point of view, cannot be considered as thoroughly satisfactory and efficient, insomuch that the great aim of sanitary science, so far as it relates to the disposal of domestic sewage, is not carried out—viz. that all sewage matters should be immediately removed to some place where they may be subjected to proper treatment without injury to health. Instead, therefore, of the fæces being removed at once they are allowed to remain within the trough or pan until such time as the whole is flushed throughout. Considering that the natural process of putrefaction or decomposition commences immediately on the evacuation of such effete organic matters as fæces, urine, &c., —especially if moisture and warmth be present—the retention of decomposing fæces within any fitment for a more or less lengthened period is not conducive to the maintenance of general health, and in the case of certain intestinal diseases may be fraught with positive danger. In addition to the sanitary objections, the use of a fitment in which the excrement previously deposited still remains, produces a feeling of disgust in all persons possessing a normal sense of personal cleanliness.

Trough closets and latrines are also difficult to keep clean, as the sides and bottom gradually become furred with fæces, except when thoroughly and frequently cleaned by hand. Unless rendered absolutely necessary for the reasons already mentioned, it is better not to use conveniences of this description, but to provide a range of independent wash-down closets of simple and inexpensive manufacture, having a separate flushing apparatus to each.

The flushing apparatus to the closets may be self-acting, the discharge being effected automatically by means of a seat or door action in situations where the users cannot be relied upon to pull the handle of the ordinary water-waste preventer.

CHAPTER XXI.

TESTING NEW DRAINS.

TESTING NEW DRAINS:—The hydrostatic or water test—Method of application—The "sweating" of drain pipes—Tested pipes recommended to be used—Testing long drains—Testing drains where embedded in concrete—Good form of drain plug—The smoke test—Description of smoke machine—Points to be observed when testing with smoke.

THE whole of the drains should be tested by hydrostatic pressure before being covered in, to make sure that every pipe and joint is perfectly sound and water-tight. Where Portland cement has been used as a jointing material, sufficient time must be allowed for it to become set before the water test is applied.

Ordinary stoneware drains, if well laid with pipes of good quality, should be able to withstand a pressure of 8 or 10 feet head of water without leakage; but it is not desirable to subject them to a greater head than that mentioned. Good cast-iron drains laid with caulked lead joints should be capable of withstanding a pressure of 200 feet head of water, but it will generally suffice if they are tested up to a working head of 10 or 20 feet.

For testing purposes, the drain is securely stopped

M

with a "drain-plug" or stopper at the lower end, and filled with water to the level of the highest gully. If additional head of water is required to that already given by the fall of the drain, a length or two of drain pipes can be temporarily fixed in a vertical position at the head of the drain, the joints being made with well-tempered clay.

The drains being filled with water to the required level (which should be carefully marked), it must be noted whether the water remains at that level for any length of time. If the same level is maintained for about two hours, the drain may be considered as satisfactorily fulfilling the test. On the other hand, if the water level is observed to be continuously falling, a systematic search for the cause of leakage must be made, every pipe and joint undergoing a rigorous examination. After the defective pipes and joints have been made good, the drain should be again filled with water, and the test re-applied until the whole is found to be perfectly water-tight.

With stoneware drains it will sometimes be seen that the water level continues *very* slowly to subside without any defective joints being discoverable. This will probably be found on examination to be due to the "sweating" of the pipes, especially if they are subject to an undue pressure of water. Notwithstanding every care being taken to use only pipes which are highly vitrified and apparently perfectly coated with an impermeable glaze, yet it is found (and may be proved by experiment) that minute quantities of water will be absorbed by and pass through the pores of the pipe. Where water has penetrated the stoneware material in this manner, the outer surface of the pipe will present an appearance similar to that of perspiration exuding from the pores of the skin. This may be reduced to a minimum by using only drain pipes which have been examined and tested before leaving the manufactory, and each pipe should bear upon it an impress of the maker's stamp to that effect.

Where the drains are too long to be tested in one operation on account of the excessive pressure that would be brought upon the pipes at the lower end, or the system is of an extensive nature, it is desirable to test the drains in sections. In cases where a section to be tested occurs between two manholes, as in a long main drain, the end discharging into the lower manhole is stopped, whilst a bend with one or more lengths of drain pipe is fixed in the manhole at the head of the section. By this means any slight leakage can easily be detected, owing to the perceptible lowering of the water level within the upright pipe; whereas, if the manhole were partially filled with water, it would require an excessive leakage to cause any noticeable difference of level over such a large area as is comprised within a manhole of normal dimensions.

The bottom of all manholes should be afterwards tested, to ascertain that the joints between the channels, drains and benchings are perfectly water-tight.

Where drains require to be embedded in concrete, or the sides of the pipes haunched up with concrete, they should first be tested and made thoroughly water-tight in themselves. Afterwards the pipes should be covered with concrete where required, and the trenches filled in and rammed whilst the drains remain filled with water, so that any damage to the pipes or joints may be immediately detected.

The form of drain-testing plug or stopper usually employed consists of a stout india-rubber ring having a

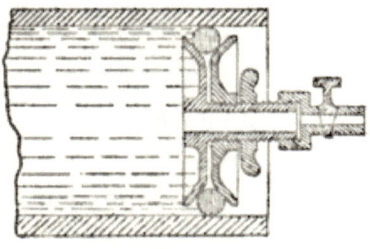

FIG. 95.

diameter slightly less than the drain to be stopped, loosely placed between two metal discs. On screwing the discs together by means of a small thumbscrew, the rubber ring is expanded or forced outwards until it is tightly compressed against the interior surfaces of the drain pipe, so as to form a perfectly air- and water-tight joint. It is desirable that the axis of the plug be in the form of a hollow pipe fitted with a screw cap or test cock, so that on completion of the test the water may be allowed to escape gradually.

Fig. 95 is a section through a drain plug of this

description, and provided with a test cock. This form of plug may consequently be used in connection with a smoke machine at any time if desired.

Another type of drain stopper consists of an india-rubber bag capable of being inflated by means of an air-pump (see Fig. 96). This form of plug is very compact and portable, and may be found useful in situations where the ordinary form of expanding plug could not be used, as for instance, when the opening to be stopped is of an oval or irregular shape.

When the gradients of the drain are so steep that

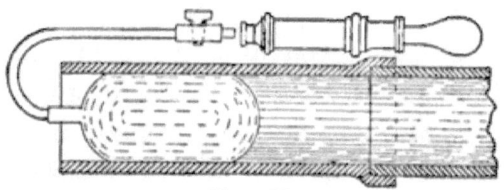

Fig. 96.

the water test would bring an excessive pressure to bear upon the lower portions, the "smoke test" may be substituted. In such cases every joint should be minutely examined whilst this test is being applied. After the smoke test has been satisfactorily carried out it is advisable to form a collar of fine concrete round each joint as an additional safeguard.

The smoke test is applied by attaching a smoke generating machine, or asphyxiator, to the lower end of the drain, the smoke being driven into it by means of a fan or bellows attached to the machine, whilst the other ends of the drain must be stopped with drain plugs

Where the drains terminate in trapped gullies, it will be sufficient seal to fill the traps with water.

The smoke test should also be applied to all soil and ventilating pipes. The asphyxiator must be connected to the foot of the soil pipe, the top of the ventilating pipe plugged, and the traps of the closets filled with water. Where it is not convenient to attach the asphyxiator to the foot of the soil pipe, the smoke may be driven into the soil pipe from the manhole into which the soil drain discharges, the machine being connected to the drain within the manhole. Whilst the soil pipes are undergoing the smoke test, particular attention should be given to all the joints that may be within the building, such as the connection of closets, housemaids' sinks, &c., to the soil pipe.

Fig. 97 shows a well-known smoke-generating machine for drain-testing purposes. It consists of a double-action bellows B connected with a copper cylinder contained within a square tank of sheet copper T. This cylinder forms the fire-box or combustion chamber, in which smoke is generated for the purposes of the test. The square copper tank surrounding it is filled with water, so that the combustion chamber may be kept as cool as possible. A deep copper cover or float is placed over the cylinder, so that the water in the tank forms a thoroughly air-tight seal or joint between them. The combustion chamber is connected to the drain to be tested with strong india-rubber tubing, and the smoke forced in by the continued working of the bellows. Large volumes of dense smoke having a powerful odour may be produced by igniting a quantity of oily cotton waste, sulphur, or prepared smoke-paper within the fire-box, and allowing it to smoulder; but precautions should be taken that the fan or bellows be carefully worked, so that the smoke material may not burst into open flame; otherwise scarcely any smoke will be obtained.

It will be found that the float of the smoke machine will rise in the water as the smoke pressure within the

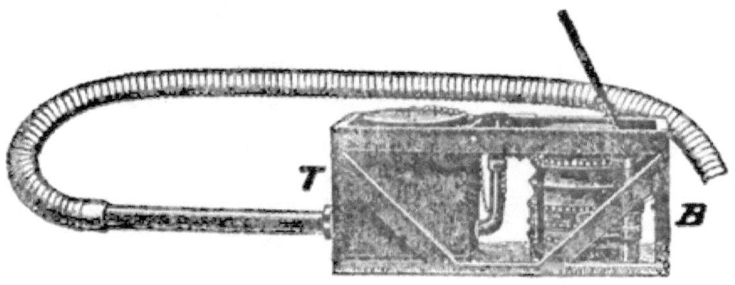

Fig. 97.

drain increases. Should the drain be thoroughly airtight, the float will remain stationary at the level to which it has been raised by the smoke pressure; but if the drain is defective the float will fall at a rate proportionate to the extent of the leakage. The action of the float in this way demonstrates the fact as to whether the drains are air-tight or not; but the leakage or leakages can only be localised by careful searching throughout the whole of the drain under test, the defects being made apparent by the issuing smoke.

CHAPTER XXII.

EXAMINING AND TESTING OLD DRAINS AND SANITARY APPLIANCES.

EXAMINING AND TESTING OLD DRAINS AND SANITARY APPLIANCES :—
Drains and fittings to be periodically examined and tested—
Drains frequently defective from unforeseen causes—Importance
of systematic and thorough examination—Method of conducting
the examination of the external drainage system—Wells and
underground tanks to be noticed—The hydrostatic and smoke
tests—Testing old drains with smoke—The examination of internal sanitary fittings—The arrangement of the water supply to
be ascertained—Smoke rockets—The scent test—Testing with oil
of peppermint.

WHILST it is essential that every drain and sanitary fitment in all its details shall be properly designed and constructed in accordance with the recognised principles of sanitary science, so that the whole may be absolutely efficient when first constructed, yet in the interests of health it is also desirable that they should be periodically examined and tested—say once a year—by a thoroughly qualified sanitary expert. Unfortunately, even in the houses of the wealthy, the instances in which such a rule is adopted and properly carried out are comparatively rare, the more usual practice being to defer such examination until it is unmistakably realised that some hitherto unnoticed defect has been the cause of serious illness or disease, and which, had it been discovered in time, might have been easily remedied.

It must be borne in mind that, under the most favourable circumstances, drains and sanitary fittings are subject to the deteriorating influence of ordinary wear and tear, and also to inevitable, if slow, decay. On these grounds alone it is necessary that they should be periodically examined and repaired if required. In addition, the drains are constantly liable to be rendered inefficient at some point through accidental or unforeseen causes, such as a settlement of the ground or building, whilst the fitments are often damaged or rendered insanitary through improper or careless use. By a careful examination from time to time any defects arising from such causes are ascertained, and may then be remedied at a trifling expense compared to that which may eventually become necessary if allowed to continue for a short period without attention.

When making an examination respecting the sanitary condition of the drains and fitments of any building, it is necessary that the whole method of procedure should be *systematic and thorough*, or the final results will be unreliable and altogether unsatisfactory. Nothing in connection with the examination should be left to guesswork or taken for granted, but everything that may affect the hygienic condition of the building under consideration should be exhaustively examined and tested. To do this effectually requires considerable experience, together with a large amount of judgment, time and patience, and is altogether a very different proceeding from the superficial operations usually carried out and considered sufficient for such a purpose.

An accurate knowledge of the general plan of the drainage system, situation of outfall, the gradients and flushing arrangements of the various branch drains, and the general character of drainage discharged by each, should first be obtained.

The position of the intercepting and inspecting chambers should be noted. Their interiors should be closely examined, and if necessary, tested, to see if the bottoms and benchings are sound and water-tight. It should also be observed if the branch or main channels have any tendency to cause splashing or obstruct in any way the even flow of sewage. The condition of the intercepting trap should be remarked—its degree of self-cleansing action and depth of water seal; also that the stopper to the cleaning arm is properly fixed.

The various gully traps should also be examined as regards their state, depth of water seal, &c., also whether they are rendered inefficient by the water seal being momentumed or siphoned out by the discharge of water into the traps or the passage of sewage through the drain with which they may be connected.

The provision of fresh air inlets and foul air outlets to the drains (also whether sufficiently adequate for the purpose), together with their position as regards the doors, windows, chimneys, &c., of the house, and the description of joint used for the foul air extracting shafts should be noted. Also that all soil pipes are carried up full bore and provided with wire guards at the top, that the rain-water pipes are entirely disconnected from the foul drains, and that the overflows from cisterns, safes, fitments, &c., discharge into the open air in suitable situations. It should be ascertained whether any drains pass under the building, and if so, the character of such drain and its construction; also whether any cesspools or neglected refuse heaps are situated near the house.

Any wells or underground rain-water tanks should be examined, and if used for drinking purposes, the water analysed. The course of the overflow from the rain-water tank should also be determined.

To ascertain whether the joints and pipes of the drains are thoroughly sound, various "tests" have been devised, the more important being those known as the "hydrostatic" or "water test," and the "smoke test." The details connected with the application of both these tests have already been given. The results obtained by the water test are the most trustworthy, though this test cannot always be conveniently applied to existing drains. Usually the smoke test is adopted; but where a drain passes under any building, it is desirable to make arrangements for that portion being thoroughly tested with water.

When testing drain, soil, or ventilating pipes with

smoke, the test should be carried out by means of a machine which is capable of *forcing* the smoke into the pipes with some degree of pressure instead of simply filling them with smoke and allowing the same to find its way through any defects unassisted.

Having ascertained that the house drains are disconnected from the public sewer by means of an intercepting trap, and that both it and all the gully traps connected with the drains are well sealed with water, the smoke machine is attached to the fresh air inlet of the drainage system, and smoke driven in until it is seen to escape freely at the outlets of all the soil and ventilating pipes. The outlets of these pipes (and any other openings connected with the drains) must then be securely plugged with a damp cloth or well-tempered clay, and the smoke forced in under pressure.

Where a smoke machine similar to that shown in Fig. 97 is used, any defects are immediately made known by the constant falling of the cylinder connected with the combustion chamber, in addition to the smoke being seen to issue through any imperfect joints or other defects. In places where it is not convenient to attach the smoke machine to the fresh air inlet, the smoke may be driven in through any gully trap that may be suitably situated for the purpose, the water seal being first removed.

It is important to notice whether the smoke issues freely from the ventilating pipes or foul air extracting shafts without any restriction whatever, in order to make sure that they are properly carrying out the object for which they were intended. The bend at the foot of a foul air extracting shaft has been frequently found completely choked with rubbish, iron rust, &c., the shaft thus being rendered utterly useless for ventilation or foul air extraction purposes. In the same way where a mica flap fresh air inlet is fixed, it should be ascertained that the inlet pipe is quite clear, more particularly at the bend, and also that the mica flap valve is working properly. When the external drains and soil pipes are being tested with smoke under pressure, it should be observed if there is any escape of smoke inside the building through any defective closet traps or joints on the branches of the soil pipes.

The examination and testing of the whole of the external drainage system having been completed, the condition and efficiency of the internal sanitary arrangements should be ascertained. It is desirable to commence at the basement or lowest floor of the building, taking the successive floors in regular progression, and carefully examining every room. All pipe casings, enclosures to baths, water-closets, &c., should first be removed, so that the whole of the fittings may be exposed to view, and the course of the various pipes traced throughout their length.

Any trapped gullies situated in the cellar or basement, and directly connected with the foul drains, should be noted for removal, as the trap is liable to become unsealed through evaporation. When cellar or basement floors require washing they should be cleaned down with water and a cloth or mop, instead of gully traps being fixed to carry off the dirty water from the floors. In cases where it is absolutely necessary to provide for waste liquids being conveyed from the cellars or basements, they should be removed by means of surface channels discharging over a trapped gully outside the building.

It should be seen if the kitchen, scullery and butler's sinks are properly trapped, and if the traps are each provided with an anti-siphonage pipe. Also that the waste pipes are disconnected from the drains and discharge over trapped gullies. Where anti-siphonage pipes are fixed, they should be examined as to whether the air way is free and not in any way blocked with dirt. If anti-siphonage pipes are not provided, the traps should be tested in order to ascertain if any of them can be untrapped by the water seal being siphoned or momentumed out.

The details of construction of each water-closet apparatus should be carefully noted, especially in the following particulars:—The volume and force of flush obtainable; the cleansing properties of the basin and trap; whether provided with anti-siphonage pipe or not; liability of water seal to be broken by siphonage or force of momentum; character of joint between trap and soil pipe; also between trap and anti-siphonage pipe; if supplied from water-waste preventer or a cistern properly disconnected from the general water supply; and whether the water-closet chamber is thoroughly ventilated. Such points as those mentioned should be observed, whatever the type of closet under examination, whether wash-down or valve pattern. In addition, valve closets should be provided with a tray or safe having a waste pipe discharging into the open air, and fitted with hinged flap valve, also an air pipe to the valve box discharging into the open air; and the closet basin should have an overflow weir discharging into the safe, or a deeply trapped overflow arm discharging into the air pipe. In the latter case it should be observed whether the overflow arm receives an adequate supply of clean water on each occasion that the closet is flushed. The valve under the closet basin must also be examined as to whether it fits closely to its seating.

All baths, lavatories, urinals, &c., should undergo a similar minute scrutiny, as already indicated for water-closet fittings. It should be seen that the wastes are trapped in every instance, and the waste pipes traced to make sure that they discharge into the open air over trapped gullies, and are not in any way connected to the drains. The character of the overflow pipes and the condition of their interiors should also be noticed.

The position and general arrangement of the domestic water supply must be carefully examined. It should be observed if the cisterns have well-ventilated surroundings, easily accessible, provided with covers, and the overflows discharging into the open air well away from any chance contamination with sewage or sewer air. The water service to the water-closets, &c., should be completely disconnected from the remainder of the domestic supply.

Small tubes containing a quantity of smoke-producing material, and known as "smoke rockets," are sometimes used for testing short lengths of drain or soil pipes. When lighted they emit a dense yellow or black smoke with pungent odour. Smoke rockets are very portable, and oftentimes convenient; but a smoke test of this description is much inferior to that afforded by a good smoke machine which forces the smoke into the drains with a slight degree of pressure.

What is known as the "scent test" or "smell test" is also occasionally used in the examination of drains and soil pipes. Certain substances having a very penetrative, distinctive and easily recognisable odour, are introduced within the drains or soil pipe at a convenient point (the outlets and drain openings having been previously plugged or sealed) so that the odour may permeate the whole system. Any defects are then made known by the sense of smell of the person making the examination. This test, though convenient under some circumstances, is not so reliable as the smoke test.

Oil of peppermint is most generally used for the scent test. This should be mixed with a quantity of boiling water and immediately discharged into the soil pipe through some conveniently placed water-closet. The mixture must be poured into the soil pipe by an assistant, and the closet afterwards well covered with damp cloths; otherwise the odour of peppermint is liable to be carried through the house and the test rendered ineffective.

Small glass phials filled with a powerfully smelling mixture may also be bought ready prepared for use in carrying out similar smell tests; but whatever the character of the test employed, it is necessary that the greatest care should be exercised, if accurate and satisfactory results are required.

INDEX.

Access to branch drains, 86
— to inspection chambers, 58
Action of sun on lead soil pipes, 64
After-flush water-waste preventing cisterns, 110
Air currents in drains, 26
— inlet valves, 29
— inlets for drains to be sufficiently large, 32
— — to drains, 28
Air-tight manhole covers, 73
Ammonia from urine, 143
Angular urinal basins, 146
Angus-Smith preservative process, 40
Anti-D traps, 80
Anti-siphonage pipe connections, 116
— pipes, 82
Arches over drains passing through walls, 52
Arrangements for flushing drains, 34
Asphyxiator for drain testing, 165
Automatic flushing chamber, 36

Base to bend at foot of soil pipe, 65
Base-plate for urinals, 146
Basin urinals, 145

Basins for lavatories, 135
Bath cradling, 130
— safes, 129
— wastes, size of, 70
Baths, materials for, 121
— size of, 123
— to be unenclosed, 129
— weight of, 123
— with secret overflow, 124
Bed of concrete under drains, 49
Bell traps, 79
Benchings, how formed, 57
Bends for open channels, 57
Blocking pieces for soil pipes, 67
Boning-staff, 47
Bower-Barff preservative process, 41
Branch drains, how connected to main drains, 54
Brass gratings to anti-siphonage pipes, 92
— — to sinks, size of, 92
Butlers' sink waste, size of, 70

Cascade action for intercepting traps, 76
Cast-iron baths, 122
— drain pipes, weight of, 42
— soil pipes, weight of, 66
— urinal basins, 149

Cast-iron ventilating pipes, 67
Cement joint for drain pipes, 44
Chamber for automatic flushing of drains, 36
Chambers, inspection, 54
— intercepting, 61
Change of direction in drains, 55
Channel bends, 57
Channels in inspection chambers, 55
Circular inspection chambers, 58
Cistern for flushing urinals, 112
— overflows to discharge in open air, 107
Cisterns, water-waste preventing, 107
Cleaning arm to branch drains, 86
Closet cisterns with after-flush, 110
— enclosures, 100
— valve with regulator, 110
Closets, hopper, 101
— pan, 95
— plunger or plug, 96
— quantity of water for, 108
— siphonic action, 104
— trapless, 97
— trough, 155
— used as urinals, 144
— valve, 98
— wash-down, 103
— wash-out, 102
Coating cast-iron pipes, 40
Combined wash-up sink and slop sink, 132
Composition of urine, 143
Concrete under drains, 49
Cone to sink waste pipes, 83
Connection between lead soil pipe and iron, 64
— between lead soil pipe and stoneware, 65

Connections for water-closets to soil pipes, 113
Conservancy system of drainage, 3
Contraction of intercepting traps, 77
Copper baths, 122
Covers for manholes, size of, 71
Cowls unnecessary to ventilating pipes, 33
Cradling for baths, 130
Creeping action of lead soil pipes, 64
Currents of air in drains, 26

D-Traps, 79
Discharge and velocity of sewage, 23
Disconnecting or intercepting traps, 75
Disposal of storm water, 6
Dr. Angus-Smith's process, 40
Double manhole cover, 73
— — — with sunk top, 73
Double-lined joint, "Hassall's," 46
Double-seal joint, 45
Drain pipe joints, 43
— pipes, earthenware, 39
— — iron, 40
— — — weight of, 42
— — stoneware, 38
— — — weight of, 40
— testing, 161, 168
— — plugs, 164
— — with oil of peppermint, 176
— ventilation, 26
Drains, concrete bed for, 49
— enlargement of, 53
— filling in trenches for, 51
— flushing chamber for, 36
— — of, 33
— foul, ventilation of, 28

INDEX.

Drains, fresh air inlets to, 28
— junctions of, 54
— laid in swampy ground, 42
— — near buildings, 51
— laying of, 47
— maximum discharge of, 24
— minimum size of, 25
— passing under buildings, 52
— storm-water, ventilation of, 28
— to be disconnected from sewer, 5
— to be outside buildings, 5
— to be self-cleansing, 13
— to be well flushed, 33
— trenches for, 50
Drainage, conservancy system, 3
— system, general plan of, 10
— water-carriage system, 3
Drawn lead traps, 81

ENAMELLED iron baths, 122
— — pipes, 41
Enclosures to water-closets, 100
Enlargement of drains, how made, 53
Examination of domestic water supply, 175
— of old drains and sanitary appliances, 168
Expanding drain plug, 164
Extraction pipes to be carried above eaves, 33
Eytelwein's formula, 18

FATTY matters from scullery sinks, treatment of, 86
Ferrules for soil pipes, 64
— for water-closet connections, 114
Field's automatic flushing apparatus, 36
Filling in trenches to drains, 51
Fire-clay baths, 121

Fixed lavatory basins, 135
Flap valves to waste pipes, 70
Flat back urinal basins, 146
Floors of bath-rooms, 129
Flow and velocity of sewage, 16
— of sewage necessary for self-cleansing drains, 20
Flushing chamber for drains, 36
— cisterns to scullery sinks, 90
— — to urinals, 112, 154
— — to water-closets, 107
— of drains, 33
— rim grease gully, 90
— — to latrine pans, 157
— — to lavatory basins, 135
— — to slop sinks, 133
— tanks to grease gullies, 90
Foot to bend at bottom of soil pipe, 65
Formula, Eytelwein's, 18
Foul drains, ventilation of, 28
Fresh air inlets to drains, 29

GALVANISED iron ventilating pipes, 67
Gases passing through water seal of trap, 76
Glass enamelled pipes, 41
Gradients for self-cleansing drains, 20
Gratings for sinks, 92
Grease from scullery sinks, 86
— gully with flushing rim, 90
— traps, 88
— — ventilation of, 88
Gullies for foul drains to be trapped, 8
— for storm-water section to be trapless, 8
— surface, trapless, 84
— — trapped, 85

HASSALL's "double-lined" joint, 46
Head of water for testing drains, 161
Holderbats for soil pipes, 67
Hopper closets, 101
House drains to be disconnected from sewer, 5
—— ventilation of, 26
Housemaids' slop sinks, 131
— wash-up sink, 131
Hydraulic mean depth, 18
Hydrostatic test for drains, 161

INDIA-RUBBER cones, 109
— drain stoppers, 165
Inlets for fresh air to be sufficiently large, 32
———— to drains, 28
Inspection arm to branch drains, 86
— chambers, 54
Intercepting chambers, 61
— traps, 75
—— details of fixing, 61
Iron baths, 122
— drain pipes, 40
——— joint for, 43
——— weight of, 42
— pipes, glass enamelled, 41
—— preservative processes for, 40
— soil pipes, 65

JOINT between soil pipe and drain, 64
— "double-seal," 45
— for manhole covers, 72, 74
— Hassall's "double-lined," 46
Joints for drain pipes, 43
— for water-closets and soil pipes, 113
Junction of drains, 54

KITCHEN sinks, 92
—— to discharge into a flushing-rim grease gully, 92
Knotting soil pipes before painting, 67

LADDER irons, 58
Latrine pans with flushing rim, 157
Latrines, 153
— to be unenclosed, 158
— water supply, 158
Lavatory basin with concealed overflow, 139
———— flushing rim, 135
———— solid waste plug, 138
———— visible overflow, 139
— basins, 135
— wastes, size of, 142
Laying drains, 47
Lead joints for iron pipes, 67
— pipes, weight of, 63, 69
— siphon traps, 80
— soil pipes, 63
— tacks for pipes, 64
Length of iron pipes, 40, 66
— of lead soil pipes, 63
— of stoneware pipes, 39, 40
Lift-up lavatory basins, 136
Long hopper closets, 101

MACHINE for generating smoke, 166
Manhole covers, size of, 71
—— solid, 72
—— ventilating, 71
—— with sunk top, 72
Manholes, inspection, 54
— intercepting, 61
— rock concrete tube, 58
Marble baths, 121
— lavatory tops, how fixed, 137

INDEX.

Maximum discharge of drain pipes, 24
Mean velocity of sewage, 16
Minimum size of drains, 25
Mica flap air inlet valves, 29
Momentum, loss of water-seal through, 81

New drains, testing of, 161

Oil of peppermint for drain testing, 176
Old drains, examination of, 168
Open channels for inspection chambers, 55
Outlets of closets to be of P form, 118
Overflow from storage cisterns, 107
— — valve closets, 99
— — wash-up sinks, 132
Oxidation of iron pipes, 67

P Outlets to traps, 118
Painting pipes coated with Angus-Smith process, 67
Pan closets, 95
Perforated gratings to sinks, 92
Pipes, earthenware, 39
— glass enamelled, 41
— iron, joint for, 43
— — preservatives for, 40
— — weight of, 42, 66
— overflow, 112
— soil, iron, weight of, 66
— — lead, weight of, 63
— stoneware, 38
— — joint for, 44

Pipes, stoneware, to be selected, 39
— — weight of, 40
— ventilating, 67
— — to be carried above eaves, 33
— waste, how fixed, 68
— — iron, weight of, 66
— — lead, weight of, 69
Plug closets, 96
Plugs for drain testing, 164
Plumbers' traps, 78
Plunger closets, 96
Porcelain baths, 121
Portland cement for jointing lead and stoneware pipes, 65
— — for jointing stoneware drains, 44
— — — — stoneware with iron soil pipes, 115
Position of anti-siphonage pipe, 82
Preservative processes for iron pipes, 40
Puff pipe to valve closets, 99

Quantity of water for flushing closets, 108
— — — for flushing drains, 34

Rainfall, usual provision for, 12
Rain-water shoes, 85
Regulator closet valves, 111
Rockets for drain testing, 176
Rusting of ventilating pipes, 67

Safe waste pipes, size of, 70
Safes for baths, 129
Salt-glazed pipes, 38
Scent test, 176

Screw caps to siphon traps, 80
Scullery sinks, 92
Secret overflow to baths, 124
Self-cleansing intercepting traps, 76
— siphon traps, 80
Self-flushing basin urinals, 149
Sewage, average daily volume, 12
— definition of, 4
— maximum hourly discharge, 12
— mean velocity of, 16
— table of velocity and discharge of, 23
— velocity and flow of, 16
Sewer disconnection, 61
Shoes, rain-water, 85
Short hopper closets, 101
Sight rails, 47
Sink waste pipes, size of, 70
Sinks, housemaids', 131
— kitchen and scullery, 92
Siphon traps, 80
Siphonic action closets, 104
Size and thickness of stoneware pipes, 40
— and weight of iron pipes, 42, 66
— — — of lead pipes, 63, 69
— of baths, 123
Slate baths, 121
Slop and washing-up sink combined, 132
— sinks, 131
— tops, 131
Smoke-generating machine, 166
— rockets, 176
— test for drains, 165
Sockets for jointing earthenware to lead soil pipes, 115
Soil pipes, 62
— — iron, size and weight of, 66
— — lead, size and weight of, 63
— — to be tested with smoke, 166
— — to slop sinks, 134

Soldered joints for lead pipes, 64
Solid manhole cover, 72
Sparge pipe to urinals, 151
Stable drains to be distinct from house drains, 12
Stall urinals, 150
— — with rounded backs, 152
Steel soil pipes, 62
Stoneware drain pipes, 38
Stoneware drains, how laid, 47
— — joints for, 44
— — maximum gradients for, 15
— pipes, to be selected, 39
— — weight and thickness of, 40
— soil pipes, unsuitability of, 62
Stopper for drain testing, 164
Storm water, definition of, 4
— — disposal of, 7
— — drains, ventilation of, 26
— — gullies to be trapless, 8
— — quantity to be removed, 12
Sunk manhole covers, 72, 73
Supply of water to latrines, 158
— pipe to valve closets, 112
Surface gullies, trapless, 85
— — trapped, 85
Sweating of drain pipes, 162

TABLE of gradients for drains, 20
— of velocity and discharge of sewage, 23
— of weights, &c., for iron pipes, 42, 66
— — — — for lead pipes, 63, 69
— — — — for stoneware pipes, 40
Tacks for lead soil pipes, 64
Taper pipes or channels, 53
Testing drains with oil of peppermint, 176
— new drains, 161

INDEX.

Testing old drains, 168
— soil and ventilating pipes, 166
Tip-up lavatory basins, 136
Trapless surface gullies, 85
— water-closets, 97
Trapped surface gullies, 85
Traps, anti-D, 80
— bell, 79
— D, 79
— for retention of grease, 88
— intercepting, 75
— — details of fixing, 61
— plumbers', 78
— siphon, 75, 80
Treadle-action flushing apparatus, 149
Trenches, filling in, 51
— for drains, 49
Trough closets, 155
— urinals, 153

Urea, 143
Uric acid, 144
Urinal basins, 145
— flushing cisterns, 112, 154
— stalls, 150
— — with rounded backs, 152
— troughs, 153
— wastes, size of, 70
Urinals, 143
Urine, composition of, 143

Valve and regulator apparatus, 111
— closets, 98
Velocity and discharge of sewage, table of, 23
— and flow of sewage, 16

Velocity of flow for self-cleansing drains, 20
— of flow necessary to remove different substances, 19
Ventilating manhole cover, 71
— pipes, 67
— — to be carried above eaves, 33
— — to be tested with smoke, 166
Ventilation of foul drains, 28
— of grease traps, 88
— of storm-water drains, 28
— of valve box to closets, 99

Wash-down closets, 103
Wash-out closets, 102
Wash-up sinks, 131
Waste pipes from bath safes, 129
— — closet safes, 99
— — lead, weight of, 69
— — size of, 70
Water, average daily consumption, 12
— companies and baths, 126
— for flushing drains, quantity required, 34
— supply to be distinct from drains, 4
— — to latrines, 158
— test for drains, 161
— waste preventing cisterns, 107
Water-carriage system of drainage, 3
Water-closet connections, 113
— enclosures, 100
— safes, 99
— supply valves, 111
— urinals, 144
Water-closets, long hopper, 101
— pan, 95
— plunger or plug, 96 :

Water-closets, short hopper, 101
— siphonic action, 104
— trapless, 97
— trough, 155
— valve, 98
— wash-down, 103
— wash-out, 102
Waterfall action for intercepting traps, 76
Water seal to intercepting traps, 75
Weight, &c., of iron pipes, 42, 66
— — of lead pipes, 63, 69
Weight, &c., of stoneware pipes, 40
— of baths, 123
Wiped lead joints, 64
Wire guard to top of ventilating pipes, 33
Wood cradling for baths, 130
Wrought-iron soil pipes, 62
— ventilating pipes, 67

Zinc baths, 122

www.ingramcontent.com/pod-product-compliance
Lightning Source LLC
Chambersburg PA
CBHW020910230426
43666CB00008B/1394